Sulfur

History, Technology, Applications & Industry

Gerald Kutney

ChemTec Publishing

Toronto 2007

Published by ChemTec Publishing
38 Earswick Drive, Toronto, Ontario M1E 1C6, Canada

© ChemTec Publishing, 2007
ISBN 978-1-895198-37-9

Cover design: Anita Wypych
Photograph (back cover) by Mitchell Kutney

Library and Archives Canadian Cataloguing in Publication

Kutney, Gerald, 1953-
    Sulfur. History, technology, applications & industry / Gerald Kutney.

Includes bibliographical references and index.
ISBN 978-1-895198-37-9

    1. Suphur, 2. Suphur industry. I. Title

QD181.S1K88 2007   553.6'68   C2007-900399-0

Printed in United States and United Kingdom

To my dear wife Denise, who patiently proof-read the manuscript with great skill and dedication.

Gerry

TABLE OF CONTENTS

Chapter 1

LE ROI DU SOL

Sulfur is the *le roi du sol* of the elements. The molecular anatomy of this sun-colored mineral is an eight-member ring that takes on the shape of a (twisted) atomic crown. Among its exceptional features is its vibrant color. The bright yellow hue is a bit of a mystery to science. While theories exist, none satisfactorily explain why crystals of this elemental crown radiate with this characteristic vibrant glow.

Another unusual property is that sulfur melts at a temperature only slightly above the boiling point of water. When liquid sulfur is cooled, prismatic ("monoclinic") crystals first form but not for long. Although it is now in a solid state, within a day, octahedral ("orthorhombic") crystals have taken its place; an observation first made by Eilhard Mitscherlich (1794 - 1863) in 1825. This inanimate, polygonal metamorphosis is a rare natural phenomenon. The two crystals have different physical properties, one of the most dramatic is that the more stable orthorhombic form has 5% greater density. The practical impact is that the solid crystal shrinks as it goes from monoclinic to orthorhombic. This collapse destabilizes the structure resulting in sulfur being friable. A hunk of pure sulfur can be easily crushed by hand. The two structures also have distinct melting points, causing confusion even among those from the sulfur world. As normal orthorhombic sulfur approaches its melting point (113°C), some of the crystals have reverted back to the monoclinic form, which melts at 119°C. Thus, sulfur will not completely melt until the latter temperature is reached.

Native sulfur is commonly found within anhydrite-gypsum deposits. Outside of the common atmospheric gases, such naturally-occurring pure elements are rare. Anhydrite and gypsum are forms of calcium sulfate (i.e., $CaSO_4$; when anhydrite comes in contact with water, it becomes gypsum), which is, itself, a sulfur-containing mineral. The most massive of these underground yellow ripples trapped within a gypsum matrix is found in Sicily, where it covers most of the center of the island. A related elemental source is anhydrite-limestone deposits found on top of salt domes. These deeper sulfur sources were first exploited on the U.S. Gulf coast. An unexpected feature is that the sulfur deposits are not created by geophysical processes. These Gaia-inspired sulfur pockets are biogenic, formed from bacterial action on the sulfate.

By coincidence, sulfur, element 16 of the periodic table, is the 16th most abundant element in nature. As with most elements, combined forms of sulfur minerals, usually sulfates (SO_4^{2-}) and sulfides (S^{-2}), are more common than elemental sulfur (S_8). From a sulfur point of view, the most important are the sulfides, in particular pyrites (iron sulfide, FeS_2; also a term used for copper sulfide). Iron pyrites can have the appearance of gold to the geological novice. The erroneous association with the other yellow element dubbed this product "fool's gold." Pyrites are not "fool's sulfur" however, for they are the most common industrial source of sulfur for sulfuric acid manufacturing outside of elemental sulfur itself. Spain was the dominant supplier of pyrites in the world. The smelting of metal sulfides, such as copper, nickel and zinc, releases sulfur dioxide, which is a source of "acid rain" if not collected. Capture of the sulfur dioxide is usually through the production of "metallurgical" sulfuric acid.

Hell is the best known "natural" (but not the most commercially attractive) source of sulfur, as popularized from the "fire and brimstone" of the Bible. The perception of purgatorial sulfur originates from its most spectacular earthly-source volcanoes, where the element is formed from the reaction of sulfur dioxide and hydrogen sulfide in the volcanic gases. Such "solfataric" sulfur is often found sputtering out of hot springs and steam vents. Solfataric mining has taken place in most volcanic areas of the world, especially in Japan, but this violent source is not significant today.

Angelic sulfur has been unfairly maligned by being associated with the work of the devil, since it is rather benign. The product is tasteless and odorless. The latter property surprises most people since the repulsive odor is well known. A recent example was recorded at the United Nations, on September 20, 2006, when the Venezuelan president, Hugo Chavez, called George Bush, the "devil himself," and to prove his point added that "it smells of sulfur still today." Such smelly sulfur does occur, not in the pure element, but in some of its simplest derivatives. The most famous of the smelly "sulfurs" are the organic sulfur chemicals that make up the concoction of the spray of the skunk. One compound of sulfur in particular is more deserving of a diabolical connotation, having the odor of rotten eggs; in fact, it is the smell of rotten eggs. The malodorous compound is hydrogen sulfide (H_2S). Oddly enough, this compound is the source of most sulfur produced today...but not from rotten eggs.

Canada is the second largest producer of elemental sulfur in the world just behind the U.S., yet a sulfur mine was never operated in this country. This apparent contradiction is the result of hydrogen sulfide being a common contaminant of natural gas in Alberta. Canadian sulfur largely originates from the purification of the gas by the removal of this hazardous chemical. Related industrial sources are petroleum, oil sands and coal, where sulfur is a component within a complex mixture of organic compounds. When sulfur is removed from petroleum, it is normally first converted to hydrogen sulfide. Sulfur derived from sour gas and oil, called "recovered," is the dominant source of sulfur today.

While *le roi du sol* has been known since before the pyramids were built, there were not many uses for it in ancient times. Arguably the first chemical element known to man, the earliest reported practical application is in the medical record called the *Ebers Papyrus*, dating from 1550 B.C., where sulfur was the active ingredient of an eye salve during the time of the Pharaohs of the early New Kingdom. The "B.C." history of sulfur, though, is virtually unknown. The only major ancient report comes relatively late from the Roman polymath, Pliny (23 A.D. - 79 A.D.). The yellow element is the topic of chapter 50, of Book 35 (p. 174 to 177) of his epic encyclopedia **Natural History**. The Roman applications were: as a medicine, in fuller's shops, bleaching wool, and in lamp wicks. Pliny called native sulfur "rhombic" (from the shape of its crystals) or "apyron" ("untouched by fire"). The major mining area of the Roman world was the Greek island of Melos, north of Crete, while pyrites (Pliny, 36.30.137) mainly came from Cyprus. Before Pliny, the ancient Greek technical literature had not much to say about this element of such great antiquity. Aristotle briefly mentions sulfur, but not in a technical sense. The earliest book on mineralogy was by his student Theophrastus (371 B.C. - 286 B.C.), called **On Stones**. He fails to discuss sulfur, as does his student, Strato of Lampsacus (fl. 290 B.C.), in his work on the Greek mining industry.

Theophrastus does write about a sulfur derivative, cinnabar (mercuric sulfide). Theophrastus described a strange reaction of this material; when cinnabar was rubbed with vinegar in a copper mortar, pure mercury was produced. This bizarre, even magical, metamorphosis is arguably the first documented chemical reaction:

$$HgS \text{ (cinnabar)} + Cu \text{ (mortar)} \rightarrow Hg \text{ (mercury)} + CuS \text{ (copper sulfide)}$$

Callias of Athens, the discoverer of cinnabar in 405 B.C., hoped that by firing cinnabar, gold could be produced. The shiny nature of this material had led him to this conclusion. The above mesmerizing transformation only reinforced the erroneous possibility that gold too could be created. Adding to the golden fallacy was that cinnabar was also known in Colchis, the land of "Golden Fleece" fame.

Pliny provides an extensive review of the cinnabar industry in his Book 33, chapters 36 to 41 (p. 85 to 95). The major source of cinnabar for the Romans was the Almaden mine (the world's oldest continuously active mine) in the Baetic region of Spain. Imports into Rome, where it was mainly used as a pigment, were one ton per year from this site.

This sulfur derivative, and subsequently sulfur as well, caught the imagination of a later age. A mismatched group of scientists, philosophers, entrepreneurs, dreamers, and swindlers, brought together by a fanatical obsession with sulfur, became collectively known as alchemists, and the yellow element became a fundamental principal of their trade. The foundations of modern chemistry were inadvertently stumbled across through their countless misguided experiments. In one of these, alchemists toying with pyrites discovered a new derivative of the sacrosanct element, sulfuric acid; the discovery, of which, is attributed to Basil Valentine (15th cent.), although the product was likely known earlier, possibly dating back to the famous Islamic alchemist Geber (~721 - ~815).

Le roi du sol remained a minor industry until industrial demand for sulfuric acid exploded in the early 19th century. Sulfuric acid became the dominant chemical in the entire world, making this era the "Sulfur Age."

Chapter 2

THE SULFUR AGE

The iconic work on the early sulfur industry is **The Stone That Burns** (and its sequel, **Brimstone, the Stone That Burns**) by Williams Haynes (1886 - 1970) in 1942. Sulfur burning is a prime candidate for the first man-made chemical reaction:

$$S \text{ (sulfur)} + O_2 \text{ (oxygen, burning)} \rightarrow SO_2 \text{ (sulfur dioxide)}$$

A clumsy, Stone-Age chemist would have discovered this reaction by accidentally dropping sulfur into a fire. The surprising result was that this yellow dirt burned with an alluring blue flame, producing a pungent odor (sulfur dioxide). Chemistry is the accidental science, as there are many more fortuitous discoveries as compared to other sciences. The most famous chemists are often the luckiest ones.

The burning of sulfur became forever linked to the element itself. Our word "sulfur" is derived from the Latin "sulfurium," which means "burning stone," as does the word "brimstone." The threatening nature of burning sulfur was well known in ancient times as documented in **Genesis** ("fire and brimstone") and other books of the Bible. Within this religious framework an ancient dichotomy arose between heathens and barbarians. While the early Christian fathers, barbarians to the Greeks, associated sulfur with hell, the earlier and more enlightened ancient Greeks, heathens to the Christians, tied it to heaven. Homer, Aristotle, and other Greeks used the term "theion," meaning "divine" and "sulfur." The Greek duality originated from the odor of sulfur dioxide being noticed from lightning, and lightning belonging to the power of Zeus. Thus burning sulfur was then connected to the chief god of Mount Olympus. The burning of this divine element was used for religious purification in Greek temples. The above chemical reaction was more than "idol" curiosity. A practical market emerged in ancient Greece from this process. Homer mentions its "pest-averting" power in the **Odyssey**, after the massacre of the suitors by Odysseus; in other words, it would stop the rapid rotting of the corpses in the hot Greek sun. The burning of sulfur as a fumigant likely originated back in Mycenaean times. By the early Roman Empire, other applications had developed, though the largest market for sulfur in the ancient world remained for fumigation and temple purification of real and supernatural pests.

An early "booming" market for sulfur involved the oxidation of sulfur to sulfur dioxide. The incendiary nature of mixtures of saltpeter, charcoal and sulfur had long been known. The chemistry behind the technology is rather mundane, when compared to the spectacular results. The saltpeter is converted into nitrogen gas and supplies oxygen for the conversion of the carbon in charcoal to carbon dioxide and the sulfur into sulfur dioxide, with the generation of much heat. The rapid expansion of these gases provides the explosive pressure to eject an object such as a cannon ball. The technology to manufacture the biggest bang, in other words gunpowder [composed of saltpeter (70%; sodium nitrate), charcoal (15%), and sulfur (15%)], was well known in China, Japan, Europe and the Ottoman Empire by the 13th century. Legend has the technology originating with the alchemist's apprentice Tu Tzu-Chhun in 762 A.D. One of the earliest battles utilizing a "bomb" was the Battle of Kaifeng, between the Chinese and Mongols, in 1232. True guns and cannons, which shot projectiles, had been invented by the early 14th century. An early, but minor, application of this weaponry was three small cannons at the Battle of Crecy in 1346. While guns and cannons were becoming commonplace among the European armies of the late 14th century, they were little more than novelty items, used for shock value rather than causing true damage to the enemy. Military strategists in Europe only realized the awesome battlefield potential of the new technology after the stunning use of huge cannons by the Turks in 1453 against the massive defenses of Constantinople, the greatest Christian fortification in the world. The first of the "weapons of mass destruction" had been discovered. Sulfur had contributed to the end of the era of the knight and his castle.

The great scientist Georgius Agricola (1494 - 1555), in his **De Natura Fossili**, wrote that he was offended by the growing use of sulfur as a weapon of war. At the end of his description of the sulfur industry, he writes that the best use of sulfur was for medicine; the:

> *"worst invention is the use...in a powder compound which, having been ignited, throws out balls of stone, iron, or copper from a new kind of military machine."*

A black market developed for the yellow element to make black powder. Until then, sulfur had been only a curiosity to the esoteric alchemists. The Church set out to limit the spread of the WMD by keeping sulfur and the other gunpowder ingredients out of the hands of the infidels. In 1527, Pope Clement VII (1478 - 1534) issued a papal bull excommunicating those who traded sulfur to "Saracens, Turks and other enemies of the Christian name." Similar decrees were issued by Pope Paul III (1468 - 1549) and Pope Urban VIII (1568 - 1644). These Papal documents are the earliest, but not the last, examples of cartel control over the international trade of sulfur.

There was not just a demand for sulfur; a nation's defense depended upon its procurement. Now, mines opened across Europe, at Bilderz in Bohemia, at Cracow, Poland (1415), in Volterra, Cesenate, and Pozzulo, Italy, at Hellin, Spain (1589) and the Hekla volcano in Iceland. Outside of Europe sulfur mines were found in Israel (Judea), Taiwan, India and Japan. While the discovery of gunpowder had produced "industrial" markets for sulfur, volumes were still limited. Even among the great military powers of Western Europe, sulfur demand was no more than a few hundred tonnes per year during the Middle Ages and Renaissance periods.

The oxidation of sulfur to sulfur dioxide, a critical step in the explosion of gunpowder, had first attracted commercial interest in the yellow element. The addition of one more atom of oxygen to sulfur would lead to industrial enterprises on a scale where no chemical had gone before. Sulfur traveled down the yellow-brick road to the fantasy land of O.V., "oil of vitriol," the alchemist terminology for sulfuric acid. So important did this foremost sulfur derivative become during the Industrial Revolution that the consumption of sulfuric acid reflected the overall economic state of a nation; the great German chemist and discoverer of synthetic fertilizers, Justus von Liebig (1803 - 1873) wrote the memorable (sulfur) statement:

> ...we may judge, with great accuracy, of the commercial prosperity of a country from the amount of sulfuric acid it consumes.

Sulfuric acid is the chemical force behind primary quintessential processes that end up creating what we call modern civilization. So, sulfur is the stealthy powerhouse of our modern world. Sulfuric acid became affectionately (at least to sulfur enthusiasts) known as the "old warhorse of chemistry." An applicable nickname for sulfur would have been Xanthus, the mythological horse that pulled Achilles' chariot into battle and is also the Greek word for yellow! Though so essential, sulfur does not usually end up in the final product. What then becomes of the sulfur? Sadly, *le roi du sol* is often discarded as a waste product, such as contaminated gypsum or dilute sulfuric acid stream. The chemical beast of burden has completed its job, and in its transformed state is no longer of any economic use.

The blossoming chemical industry of the day was driving economic growth, and this industry depended upon sulfuric acid. Although still a popular adage among sulfur aficionados, the direct connection to national prosperity is a bit of an exaggeration today. Since half of the sulfuric acid in the world is consumed to manufacture phosphate fertilizers, sulfuric acid consumption can be driven simply by the availability of phosphate rock deposits to produce fertilizers rather than general economic prosperity. That said, no other single chemical directly reflects the global economy better than sulfur consumption even today. David Morse of the U.S. Bureau Mines described the modern importance of this valuable element (**Minerals Yearbook** 1990):

> *Sulfur, through its major derivative sulfuric acid, ranks as one of the most important elements utilized by humanity as an industrial raw material.*

The cultural development of mankind has passed through many "chemical" ages. Earlier fabulous times were described as a Golden Age, followed by the Silver Age. In ancient times, there was a Copper Age, Bronze Age and an Iron Age. Terminology changed, so that the 18th and 19th centuries became known as the Industrial Age. Keeping with the chemical tradition, the more proper name should have been the "Sulfur Age," for without sulfuric acid it would never have happened.

2.1 SULFURIC ACID MANUFACTURING

2.1.1 THE NORDHAUSEN PROCESS

The first commercial production of sulfuric acid took place at Nordhausen, in Saxony, Germany, during the early 16th century and continued until 1900. Production, though, was limited in the first few centuries, with throughput being measured in only ounces. The low production reflected the market in the Middle Ages, where the major application was to clean and treat metals. Production increased, reaching a high of just less than 4,000 tonnes in 1884. While normal acid was less expensive, Nordhausen acid was selling for $90 per tonne in this peak year. Nordhausen was the most famous site; other plants existed in the Middle Ages, including one in Britain.

The acid was produced from the dry distillation of ferrous sulfate, by the following process:

$$FeSO_4 \text{ (ferrous sulfate) + heat} \rightarrow Fe_2(SO_4)_3 \text{ (ferric sulfate)}$$

$$Fe_2(SO_4)_3 + 2FeSO_4 + \text{heat} \rightarrow 4SO_3 + SO_2 + 2Fe_2O_3$$

$$SO_3 \text{ (sulfur trioxide) } + H_2SO_4 \rightarrow H_2S_2O_7 \text{ (Nordhausen acid)}$$

During the Middle Ages, pyrites were called vitriol. The alchemical lingo for ferrous sulfate was called "green vitriol," while cupric sulfate was called "blue vitriol." Sulfuric acid, itself, being produced from these products, was first known as "oil of vitriol." Another name for such "fuming acid" was "Nordhausen acid." This material is a special case of sulfuric acid, where almost no water is present. Whereas normal sulfuric acid is sulfur trioxide dissolved in water, fuming acid is sulfur trioxide dissolved in concentrated sulfuric acid.

2.1.2 THE BELL PROCESS

The use of elemental sulfur to produce sulfuric acid came later. Reacting elemental sulfur with oxygen (i.e., burning) stopped at sulfur dioxide. The process had been studied by some of the leading scientists of the 16th century, Angelo Sala (1576 - 1637), Andreas Libavius (1560 - 1616), and Conrad Gesner (1516 - 1565). To add the essential one more atom of oxygen to form sulfur trioxide, a catalyst was required.

$$2SO_2 + O_2 \text{ (oxygen, catalyst)} \rightarrow 2SO_3$$

$$SO_3 + H_2O \text{ (water)} \rightarrow H_2SO_4 \text{ (sulfuric acid)}$$

Credit for this innovation is attributed to Johann Rudolf Glauber (1604 - 1670) in 1651, which is the earliest known reference to saltpeter (nitre), the other essential ingredient of gunpowder, as such a catalyst. Glauber's discovery may have been anticipated by the legendary inventor (noted for his work on microscopes and the submarine) Cornelius Drebbel (1572 - 1633), who has been claimed to have built a plant to manufacture sulfuric acid, but there is no proof of this.

The new product was amazing, yet frightening, to the earliest investigators. No one had ever seen such a corrosive material. This super acid ate through all the standard metals of the day. The only substance to cage the burning liquid was glass. While common in the laboratory, industrial-scale glass reactors ("glass bells," named because of their original shape, later replaced by globes) were expensive, cumbersome and prone to breakage.

Plants were built in France and Holland by the end of the 17th century with the Glauber technology. One of the largest plants to manufacture sulfuric acid from elemental sulfur by the Bell process was the Great Vitriol Works at Twickenham. The Twickenham plant was owned by Joshua ("Spot") Ward (1685 - 1761), a notorious figure who was known for his quack elixirs, potions and miracle medicines, and John White, in 1736. They filed a patent on their process (No. 644), which was issued on June 23, 1749. The fragile glass reactors at the Twickenham plant had a capacity of up to fifty gallons each (the limit of the glassblower's trade). In 1740, the plant was moved to Richmond, after complaints from neighbors of the smell. Ward's plants had one of the largest capacities in the world at that time, although still on the scale of only pounds. With this new large-scale Ward plant, the price of acid dropped from £3,500 per tonne (although never sold in this quantity) to £220 per tonne. This technology, though, would not last much longer, as the annoying glass bells were soon replaced by a more practical reactor. With the new technology, production was now measured in tonne quantities.

2.1.3 THE CHAMBER PROCESS

A creative breakthrough for the industrial production of sulfuric acid was made by John Roebuck (1718 - 1794) a decade later. He had learned chemistry from the famous Joseph Black (1728 - 1799) at the University of Edinburgh. He is best known as being the financial backer of James Watt (1736 - 1819) in the development of the steam engine. In 1746, Roebuck replaced the glass bells, with lead-walled reactors (i.e., chambers), increasing capacity and reducing manufacturing costs by 90%. The idea to use lead had come to Roebuck after searching the early records of Glauber, who showed that the acid did not react with this material. The first lead Chamber plant was installed at the Birmingham Vitriol Manufactory, owned by Roebuck and Samuel Garbett (1715 - 1803). However, Ward claimed that the new plant was using technology covered under his patent (only issued in England and Wales). So Roebuck and Garbett left England and went to Scotland in 1749, where they established the famous Prestonpans Vitriol Manufactory, outside of Edinburgh. Prestonpans became one of the largest chemical centers of the early Industrial revolution (the chemical operations were closed by 1825). By the end of the century, the total capacity was over 60,000 cubic feet. Their product, "English vitriol," became known across the continent. Roebuck's later investments went poorly for him and he filed for bankruptcy in 1773.

Roebuck had attempted to keep his proprietary process a trade secret, but the innovation was basically too straight-forward for such a strategy to work for long. In 1756, an ex-employee of Roebuck had convinced a Mr. Rhodes to build a similar plant at Bridgnorth. A similar situation took place with Samuel Skey (1726 - 1800) with the building of a sulfuric acid plant at Dowles. The business was carried on by his son, Samuel Skey (1759 - 1806). Other early Chamber facilities in the U.K. were:

Kingscote & Walker – 1772 – London
Thomas Farmer & Company – 1778 – London
Baker, Walker (nephew to above), Singleton – 1783 – Pilsworth Moor
Bealy, Radcliffe – 1791 – Manchester
Vitriol and Aquafortis Works – Rowson and Benjamin – 1792
(1750) –Bradford
St. Rollox Works – Charles Tennant (1768 - 1838) – 1803 – Glasgow
Doubleday & Easterby – 1809 – Bill Quay

By 1820, twenty-three Chamber plants were operating in the U.K., seven in the city of London alone.

Pricing dropped to £41 per tonne with Roebuck's operations, which opened up new markets for the now inexpensive reagent. The new competition caused sulfuric acid pricing to tumble even further. By 1798, it was £29 per tonne and then £17 per tonne in 1806. By 1815, acid pricing had recovered to £34 per tonne. English sulfuric acid production reached 2,700 tonnes in this year.

At the end of the 18th century, the production costs of sulfuric acid were down to £20 per tonne. Cost of production data for two sites differ in detail but overall production costs are similar (see Table 2.1). The long established Preston-pans site had obviously negotiated much better delivered sulfur (and nitre) costs to their site. Their delivered sulfur costs were less than half of Bealy, Radcliffe. The proportionally huge "other expenses" at Prestonpans are not explained. At Bealy, Radcliffe, there were six chambers, 12' x 10' x 10', and while at Prestonpans, there were 108 chambers, 14' x 4' x 10'.

Table 2.1. Sulfuric acid – U.K in 1800 (cost of production per tonne for commercial grade O.V.)

Cost type	Bealy, Radcliffe		Prestonpans	
	Raw material, unit cost, £	Variable cost, £	Raw material, unit cost, £	Variable cost, £
Sulfur	16.90	11.60	7.70	6.20
Nitre	70.60	6.90	39.70	4.10
Other expenses		2.60		10.40
Total cost		21.10		20.70

Source: H.W. Dickenson

Production costs soon dropped again through technology improvements. In 1838, a method of welding large lead sheets was discovered by Debassaynes de Richemond that allowed for the manufacture of larger-scale reactors. With this latest advancement, one of the largest plants in the world, at a capacity of one million cubic feet, was built at the St. Rollox works of Charles Tennant & Company, in 1861, then under the direction of Charles Tennant (1823 - 1906), the grandson of the founder of the company. By then, sulfuric acid had become a world-scale industry and Britain was the largest producer in the world.

England soon lost its monopoly on the Roebuck technology. Not only did competing plants arise in his own country, but also soon afterwards in Continental Europe. John Holker (1719 - 1786) brought the technology over to France, building a plant at Rouen in 1766, followed by another in the same city by A. De La Follie in 1774. By the end of the century, four Chamber plants were operating at the Starck Works, owned by Johann Starck, in Bohemia. In Germany, plants were built by Jean Angelo Grasselli at Mannheim in 1805 and by the Giulini Brothers (became BK Giulini Chemie; taken over by Israel Chemicals Limited in 1977) in 1823. Both German firms were importing sulfur from Sicily. The opening up of the Leblanc process (see below) led to a proliferation of the building of Chamber sulfuric acid plants. Soon the technology was commonplace among the industrialized nations of Western Europe.

The first sulfuric acid production in North America was by John Harrison (1773 - 1833) of Philadelphia in 1793, in his drug store. Although an unlikely site for chemical production today, the pharmacist's "store" was often the site of the early chemical industry. Harrison had studied chemistry under the noted chemist Joseph Priestley (1733 - 1804), who moved to Philadelphia in 1794. Harrison's laboratory-scale plant had a capacity of 20 tonnes per year by 1806. He opened a larger facility the following year, with a capacity of 250 tonnes per year. Harrison claimed that his capacity was greater than total U.S. demand. The new Chamber unit was fifty feet long and eighteen feet high. Price of the acid at the time was $330 per tonne in the U.S. In November 1808, he proudly wrote to President Thomas Jefferson describing his pioneering chemical business.

Harrison was also the first to install a new technology to produce concentrated acid. The standard Contact acid was too weak for some applications which required stronger acid but not necessarily fuming. At high concentration, even lead reacted with sulfuric acid. A new distilling technology used platinum stills, a process developed by Erick Bollman, a friend of Lafayette, in 1813. Platinum was not the exotic, expensive element of today. Bollman had found one of the few uses for this material. Platinum was relatively difficult to manipulate and a method had only recently been developed by William Hyde Wollaston (1766 - 1828), whom Bollman knew. Harrison used the original still until 1828.

The Harrison firm was a family affair, beginning as John Harrison & Sons, and under his sons, Thomas Harrison (1805 - 1900), Michael Leib Harrison (1807 - 1881), and George Leib Harrison (1811 - 1885), became the Harrison Brothers & Company. Later members of the firm were John Skelton Harrison, Thomas Skelton Harrison (? - 1919), who later became Consul General in Cairo, and George Leib Harrison (~1836 - 1935), son of Thomas. By the time of the Civil War, they had production facilities also in New York and Maryland; the lead chambers were now 160' x 26' x 24'. Total sales of the company were more than $1 million per year. More than a century after Thomas Harrison produced his first sulfuric acid, in 1918, the company was acquired by DuPont for $5.7 million.

Another early sulfuric acid plant in the Philadelphia area was built in 1831 at the Tacony Chemical Works by Nicolas Lennig & Company. Nicolas Lennig (? - 1835), founding the firm in 1819, had been joined by his son Charles (1809 -

1891). Lennig also concentrated sulfuric acid using platinum stills but on a larger scale than Harrison. Upon the death of his father, Charles was joined in the business with his cousin Frederick Lennig (? - 1863). Before the Civil War, Lennig was one of the largest chemical companies in the nation. The name of the company became the Charles Lennig Chemical Company. Later, John B. Lennig (~1851 - ?), the son of Charles, ran the company. Upon his death the firm was sold to Rohm & Haas in 1920.

Philadelphia became the center of the early American sulfuric acid industry. By the 1850's, several firms were manufacturing acid in the city and environs:

- Harrison Brothers & Company
- Charles Lennig Chemical Company
- Powers & Weightman – originally opened by Abraham Kunzi and John-Farr (? - 1847) in 1818; after the death of Kunzi, Thomas H. Powers (~1813 - 1878) and William Weightman (1813 - 1904), the nephew of Farr, became partners in 1838; the firm became the largest pharmaceutical company in the U.S., which was taken over by Merck in 1927.
- Potts & Klett – a phosphate fertilizer manufacturer, near Camden, founded by Robert Barnhill Potts (1816 - 1865) and Frederick Klett (1827 - 1869).
- Moro Phillips & Company – Aramingo Chemical Works; a phosphate fertilizer company founded by Moro Phillips (~1815 - 1885; his Polish name was Philip Chariotsky Moro), who had emigrated from Poland; later became a founding member of General Chemical (see below).
- Savage & Martin – Frankford Chemical Works; later becoming Savage & Stewart.

Philadelphian acid production was 7,700 tonnes per year at this time.

The first non-Philadelphia Chamber plant in the U.S. was built in Baltimore, MD, by William T. Davison (~1804 - 1881) in 1832. Davison Chemical Company (acquired by W.R. Grace in 1954) was an early manufacturer of phosphate fertilizers. Another major acid producer was Merrimac Chemical of Boston. The company had been formed in 1863 from the merger of Merrimac Manufacturing, founded by Patrick Tracy Jackson (1780 - 1847) in 1821, and Woburn Chemical, founded by Robert Eaton in 1853. This enterprise was purchased by Monsanto in 1929.

A later industrial leader in sulfuric acid traces its roots to New York City back in 1866. Charles W. Walter (? - 1875) and August Baumgarten (~1827 -?) had opened up a sulfuric acid plant at the Laurel Hill Chemical Works. Five years later, the business was purchased by George Henry Nichols (1823 - 1900) and his son William Henry Nichols (1852 - 1930). A larger facility was built at Laurel Hill in 1870 (the site later became the Nichols Copper Works, which was purchased by Phelps Dodge in 1930). The firm went through a number of name changes:

1866	Walter & Baumgarten
1871	Walter & Nichols
1875	G.H. Nichols & Company
1897	Nichols Chemical Company

While the early sulfuric action was naturally in the N.E. U.S., the industry spread across America. In 1839, Eugene Ramiro Grasselli (1810 - 1882) opened the first Chamber plant in the mid-west (Grasselli Chemical Company), in Cincinnati, OH. He was born in Strasburg, the son of Jean Angelo Grasselli, an early pioneer of the Leblanc process in Germany. Before coming to Cincinnati, Eugene had gotten his sulfuric acid feet wet (or burnt!) working for a chemical company in Philadelphia. Grasselli moved his head office to Cleveland and opened another plant here in 1868 to be closer to the growing petroleum industry, one of the largest users of acid in the U.S. at the time. Other Chamber plants were built by Grasselli in New Jersey and Alabama. His son Caesar Augustine Grasselli (1850 - 1927) became president in 1885, and then his son Thomas S. Grasselli (1874 - 1942) took over the firm in 1916. The company was purchased by DuPont in 1928.

Out West, the San Francisco Chemical Works opened a Chamber plant in the latter 1860's. The company was founded by Egbert Putnam Judson (1812 - 1893) and John L. N. Shepard. The site was destroyed by an explosion at the nearby Giant Powder Company, in 1892. During the 1880's, another sulfuric acid plant opened in the city by the California Chemical Works, founded by John Reynolds. These two companies, along with the operations of John Wheeler and John Stauffer (~1862 - 1940), merged to form the San Francisco Sulfur Company in 1894. The following year, the company became the Stauffer Chemical Company (acquired by Rhone-Poulenc, now Rhodia, in 1985). During the great earthquake of 1906, the plants were badly damaged and had to be rebuilt.

In Canada, a small sulfuric acid plant was opened by an obscure individual named Adams, in Saint-Jean-sur-Richelieu, QC, in the latter half of the 1860's. The source of sulfur was pyrites from Capelton. In Canada, the two early centers of sulfuric acid production were Capelton, QC, where a major pyrites deposit had been found, and at London, ON, where the fledgling Canadian petroleum industry was getting under way. The first major acid plant, the Canada Chemical Company, was opened on May 21, 1867, in London, ON, by William Bowman of Liverpool, England, and Thomas Henry Smallman, a later founder of Imperial Oil in 1880 and the London Life Insurance Company in 1874. A report on the firm from 1887 stated that the acid was produced "from pure brimstone, using platinum distilling apparatus." Sulfur from Sicily was used as the raw material. In an attempt to reduce costs, the expensive Sicilian sulfur was replaced by Canadian pyrites. However, they could not get the process to operate effectively and they returned to elemental sulfur. The Canada Chemical Company remained a leading acid manufacturer in the last quarter of the century. In 1878, they purchased the acid facility of another early producer, the Western Canada Oil Lands Limited, which had opened in London in 1870 (and was dismantled in 1888). The Canada Chemical Company was sold to the Nichols Chemical Company in 1898, and the original London site was closed in 1904. Another early London facility was the Ontario Chemical Works of MacHattie & Co. (their advertisements appeared in the *Globe* in 1868; a survey of London businesses from 1878 does not mention this company).

During the 19th century, Chamber acid plants were built around the world. As the century was closing, a new process was under development, which would become the dominant acid process of the 20th century.

2.1.4 THE CONTACT PROCESS

The major limitation of the Chamber process was the strength of the acid it produced, even after distillation. Acid from the Chamber process could be concentrated, using platinum stills, but only up to 78%. A special feature of the new Contact process was the ability to make fuming acid; still basically the monopoly of the very expensive and limited Nordhausen acid.

This technological improvement was based on a modification of the chemical process of Glauber, which had been largely unchanged for centuries. While the basic reaction remained the same, a new catalyst, platinum had been discovered (that this material was also appropriate for distilling weak acid is purely coincidental). The use of platinum as a catalyst in the process had been discovered by Johann Wolfgang Dobereiner (1780 - 1849) during the 1820's, but his work had not attracted much attention, and on March 25, 1831, a patent (No. 6096) for producing acid with a platinum catalyst was issued to the largely unknown Peregrine Phillips of Bristol. While of academic interest, the technology was not yet ready for commercialization, and since fuming acid markets were limited, there was not much incentive to overcome the practical handicaps of this new process.

The demand for such stronger acid had been small, until the emergence of the dyestuffs industry in the 1870's. Then, Rudolf Messel (1848 - 1920) in London, and Clemens Alexander Winkler (1838 - 1904) in Freiberg, worked on the process. A key part of their invention was the use of powdered platinum as the catalyst. The German-born Messel had been a chemist in the firm Squire, Chapman, and Company from 1875 to 1878. The improved, yet still flawed, technology was first installed at their sulfuric acid plant at Silvertown in 1876. In 1878, Messel became a partner in the renamed firm Spencer, Chapman and Messel, where he was Managing Director until 1916. In 1875, developments by Winkler had led to another Contact plant to be opened in Frieberg. While the new plants were competitive versus the expensive Nordhausen acid, the process was still costly and difficult to operate. Impurities in the system, especially arsenic, poisoned the catalyst. Except for a few special cases, the Contact process was still only a technical curiosity.

Secret developments in Germany by BASF overcame many of the practical problems by 1889, but the information was not released until ten years later, when the engineering modifications of Rudolf T.J. Knietsch (1854 - 1906) were patented. Knietsch presented his work to the German Chemical Society in 1901 (Berichte 34, 4069, 1901). The technology monopoly afforded by the patent restricted the early growth of this process to BASF sites. In Germany, production was over 100,000 tonnes per year by the end of the century (see Table 2.2a). In 1899, the first of the new Contact acid plants in the U.S. was built at Mineral Point, WI, followed by another in 1910 at New Jersey Zinc (now part of Sun Capital Partners, under the name of Horsehead Corporation) in Wisconsin.

Table 2.2a. Sulfuric acid – world (production in 1000 tonnes; 100% acid)

Year	Britain	France	Germany	US	Global
1867	155	125	75	40	**500**
1878	600	200	112	180	**1,300**
1900	1,000	500	950	940	**4,200**
1910	1,000	500	1,250	1,200	**5,000**
1913	1,100	n.a.	1,700	2,500	**8,500**
1925	848	1,500	1,125	4,257	**10,563**
1950	1,832	1,215	1,746	11,829	**25,700**

Source: Haynes: **The Stone That Burns**, and others

BASF continued to work in this area. In 1915, they discovered a new type of catalyst. Vanadium catalysts are less prone to poisoning, a problem with the platinum-based ones. The technology was introduced into the U.S. in 1926, and afterwards became the common technology for sulfuric acid production. Another development came from another German chemical giant, Bayer. In 1963, they introduced the Double-Contact process. The improved process has efficiencies approaching 100%, which virtually eliminates any sulfur dioxide emissions. The basic process had been suggested by General Chemical (U.S. Patent 1,789,460) in 1931, but was not commercialized.

The initial secrecy backfired for BASF in North America. John Brown Francis Herreshoff (1850 - 1932), the vice-president of the Nichols Chemical Company, independently discovered a similar process. For many years, BASF and the General Chemical Company were in litigation over the patent rights to the invention, which resulted in the General Chemical Company obtaining an exclusive license for the technology from a reluctant BASF in North America. The Contact process of BASF had, in fact, been the "catalyst" for the merger of the Chamber process companies that formed the American sulfuric acid giant General Chemical. On February 15, 1899, a major consolidation of the acid industry took place when William Nichols and his son Charles Walter Nichols (1876 - 1963) led the amalgamation of twelve sulfuric acid manufacturers into the General Chemical Company (in 1920, another amalgamation with four other companies produced Allied Chemical [later Allied Signal]): Chappel Chemical Company, Dundee Chemical Company, Fairfield Chemical Company, Highlands Chemical Company, Jas. Irwin & Company, Jas. L. Morgan & Company, Lodi Chemical Company, Martin Kalbfleisch Company, Moro Phillips & Company, National Chemical Company, Nichols Chemical Company, and Passaic Chemical Company. Nichols had gathered the competitors together with the rallying cry that their Chamber process plants were no longer competitive against the exciting new German technology. The antagonism and mistrust among this group melted as General Chemical offered them the only potential access to the technology. The elder Nichols, who had been president of the Nichols Chemical Company since 1890, became the new company's first president. He was president until 1907, and then chairman until 1920.

After settling the court case, General Chemical quickly expanded with Contact technology. At the beginning of World War I, they already had 44 Contact units at ten sites, increasing to 62 units by the end of the war in the U.S., and another six in Canada. In the U.S., the Contact process had surpassed the Chamber process as the dominant sulfuric acid technology in the mid-1930's (see Table 2.2b).

Table 2.2b. Sulfuric acid – US

Year	Plants [number]	Raw materials, %		Technology, 1000 tonnes, 100% acid		
		sulfur	pyrites	chamber	contact	total
1865				34	0	34
1867				40	0	40
1870				59	0	59
1875				75	0	75
1878				180	0	180
1880	50			240	0	240
1885		85	14	339	0	339

Table 2.2b. Sulfuric acid – US

Year	Plants [number]	Raw materials, %		Technology, 1000 tonnes, 100% acid		
		sulfur	pyrites	chamber	contact	total
1889				442	0	442
1891		80	19		0	
1895		75	24		0	
1899						874
1900						940
1901		16	81			
1904						1,055
1905		10	79			
1909		2	84			1,551
1910		3	79			1,200
1913				1,664	330	1,994
1914		3	77			2,284
1915		9	64			2,289
1916						3,514
1917				2,294	1,072	3,366
1919	216	48	36	2,120	644	3,133
1920		52	23			
1921	197			1,848	622	2,466
1923	185			2,675	1,022	3,700
1925	177	68	19	2,794	1,157	3,953
1927	181			2,884	1,254	4,140
1929	170	69	17	3,055	1,735	4,792
1930		68	14			
1931	175			2,052	1,381	3,434
1934		67	22			
1935	155	59	29	1,693	1,935	3,629
1937	157			2,338	2,146	4,482
1939	183			1,924	2,426	4,350
1940		64	24			
1941				2,732	3,409	6,142
1943				2,856	4,266	7,659
1945	174	69	12	2,875	5,006	8,638
1947	180			2,971	6,195	9,780

Table 2.2b. Sulfuric acid – US

Year	Plants [number]	Raw materials, %		Technology, 1000 tonnes, 100% acid		
		sulfur	pyrites	chamber	contact	total
1949	191			2,490	7,241	10,371
1950		74	11			11,829
1955		72	10			14,746

Source: Haynes: **Brimstone The Stone That Burns**

With General Chemical controlling the Nichols' operations in Canada, the new technology was transferred to this country as well. In 1905, the first new Contact process plant in Canada was installed at Capelton, and three years later, the process was installed at the new Nichols' plants at Sulphide (named after the local pyrites mine that had opened in 1903), ON, and Barnet, BC. Nichols had opened its first Chamber plant in Canada back in 1887, at Capelton, QC. After fire had destroyed the facility at Capelton on November 24, 1924, production was transferred to Sulphide. In 1941, Nichols opened another sulfuric acid plant at Valleyfield, QC, which was supplied pyrites from Noranda, QC. When the Canadian patents expired in 1922, other companies opened Contact plants. Among the first was the Grasselli Chemical Company at Hamilton, ON, in 1922 (the site later became part of CIL and then ICI Canada); a Chamber plant had existed since 1912. The sulfur source for Hamilton had been pyrites from Peterborough, ON, until 1920, when they switched to elemental sulfur.

2.1.5 METALLURGICAL SULFURIC ACID

The chemical innovation behind the Chamber and Contact processes was not the burning of sulfur (or pyrites) into sulfur dioxide, but to convert sulfur dioxide into sulfuric acid. These two processes became part of the earliest pollution control technologies for smelters which were emitting huge volumes of sulfur dioxide in their stack gases from the roasting of pyrites and other sulfide minerals. To reduce these acid rain emissions, the sulfur dioxide is concentrated and then passed through the Contact (or Chamber) process to manufacture sulfuric acid, usually called metallurgical acid. Recovery of sulfuric acid from smelter gases had first taken place in Britain in 1865. In the U.S., Matthiesson and Hageler (from zinc processing) in LaSalle, IL, in 1895, and the Tennessee Copper Company by 1905 were also producing metallurgical sulfuric acid from the Chamber process. In 1911, metallurgical sulfur acid production in the U.S. surpassed 100,000 tonnes (based on sulfur equivalents) for the first time.

In 1925, the first major recovered sulfur dioxide-to-sulfuric acid plant in Canada had been opened by Mond Nickel (merged with INCO in 1929) in Coniston (closed 1930), outside of Sudbury, ON. The experience from Coniston was utilized to open a second plant (Contact process) at Copper Cliff, ON, in 1930 by INCO with CIL (then a joint venture between ICI and DuPont). Earlier, a small unit had been installed by Cominco at Trail, BC, in 1916, using the Chamber process. A full-scale unit (Contact process) followed in 1931 (after another small unit had opened two years earlier). The decision to build the larger plant was driven by legal claims from Washington-state farmers that their crops were being damaged by the sulfur dioxide fumes of the smelter. Cominco also opened an ammonium sulfate fertilizer plant at nearby Warfield Flat, BC, to use the sulfuric acid. Another sulfuric acid plant was brought on stream in Kimberley, BC, by Cominco. During the 1940's, metallurgical acid production in Canada was over 100,000 tonnes per year.

The clear viscous liquid sulfuric acid and yellow solid sulfur have become inseparable, at least within the statistics of global data collection. Metallurgical sulfuric acid and pyrites used for sulfuric acid manufacturing are considered "equivalent sulfur;" they are included as sulfur statistics by the U.S. Geological Survey and National Resources Canada. Sulfuric acid from pyrites appears under two separate sulfur categories, and there is but a subtle difference between them. If a smelter roasts pyrites and produces by-product sulfuric acid, it is labeled metallurgical sulfuric acid. If, on the other hand, pyrites are sold to a third party for sulfuric acid manufacturing, then they are labeled simply pyrites. Both are included in overall ("sulfur in all forms") sulfur statistics. Consequently, there are several categories of "sulfur:" native (i.e., mined sulfur, other than Frasch), Frasch (see below), and oil & gas (these three categories are all elemental sulfur), and pyrites and metallurgy (the latter two categories are normally sulfuric acid but small amounts have been converted to elemental sulfur). The synonymic relationship between sulfur and sulfuric acid goes beyond the direct connection of most sulfur being converted to the latter. Sulfuric acid is hazardous to transport, while sulfur is not. Sulfur can be thought of as a docile, concentrated facsimile for sulfuric acid, a sort of "instant acid." The world trade in sulfur is driven by sulfuric acid, not sulfur per se.

2.1.6 PYRITES

Pyrites, too, are reported by the U.S. Geological Survey and National Resources Canada, in sulfur statistics. The substitute product for sulfur in the manufacture of sulfuric acid was pyrites (note: this is the singular and plural spelling of the word). For most of the first half of the 19th century, pyrites had little market share, but during the second half, pyrites dominated the sulfuric acid market.

Roasted pyrites release sulfur dioxide, which can then be converted to sulfuric acid. Commercial scale production of sulfuric acid from pyrites had taken place back in 1793, by M. Dartigues of France. To produce sulfuric acid from pyrites, the critical technology was the pyrites burner. A major technical break-through was an improved furnace, invented by Michel Perret (1813 - 1900) in 1832 (Fr. Patent No. 1094, 1836; another modification by Perret was made in 1845). Perret built a sulfuric acid plant in Saint-Fons, near Lyons, in 1837. Later burner developments came from MacDougall Bros. of Liverpool and James Brown Herreshoff Sr. (1831 - 1930), brother of John Brown Herreshoff.

In Britain, Thomas Hills and Uriah Haddock used pyrites at Bromley-by-Bow, but only for a short period. A patent on their process appeared in 1818 (No. 4263). The pyrites process was utilized on a larger scale by Thomas Farmer & Company at Kennington in 1838, who had been one of the earliest entrants into the Chamber acid business in 1778. Pyrites had only come into their own as a serious sulfur source for sulfuric acid manufacturing after the TAC episode and the associated Sulfur War (see Chapter 3).

After the technology switch of 1840 to pyrites, Ireland was the initial supplier of pyrites into Britain; the first shipment taking place on December 28, 1839, from the Ballymurtagh mine of the Wicklow Copper-Mine Company (opened in 1822). The pyrites went to the firm of Newton, Keats and Company of Liverpool. During the 1830's, this mine had only been producing 3,000 to 5,000 tonnes per year of low-grade pyrites. After the TAC incident, sales of the company rose to 15,000 to 20,000 tonnes per year and total Irish pyrites exports averaged 40,000 tonnes per year in the 1840's, jumping to 100,000 tonnes per year early in the following decade. In the 1860's, Wicklow opened its own sulfuric acid plant; the acid was used to manufacture phosphate fertilizers. Pyrites were also shipped from Cornwall. In 1840, more than 6,000 tonnes were shipped from here to the Leblanc producers.

Major pyrites deposits were also developed in Japan, Canada, the United States, and other countries by the middle of the 19th century. Out of the "fool's gold rush" to take advantage of the rising demand, the output of one country soon outpaced those of all other countries combined; that country was Spain.

2.1.6.1 Spain

Spanish pyrites were first imported into Britain from the Tharsis region of Spain by the Tharsis Sulphur and Copper Company (TSCC) founded in 1867 by the Scottish industrialist John Tennant and his son Charles. Charles was chairman of Tharsis from 1867 to 1906.

More rich pyrites resources lay in Southern Spain at Rio Tinto. These assets had been under the control of the crown until the Spanish government put the resources up for sale in June 1870. The government asked for the huge sum of $4 million for the under-developed mines, but received no takers. The highest bidder was the Scottish businessman Hugh Matheson (~1820 - 1898) and his syndicate. Matheson worked for Jardine-Matheson, which had been founded by William Jar-

dine and James Sutherland Matheson (his uncle) in 1833. As Matheson was raising money for his acquisition, Charles Tennant and TSCC fought to derail the
purchase; he publicly claimed that their prospectus was fraudulent. Matheson per-
sisted and raised his funding. A major backer of Matheson was the Rothschild
family, through N.M. Rothschild and Sons. The main commercial interest in Rio
Tinto was not for sulfur supply for sulfuric acid manufacturing, but for copper
production. Rio Tinto soon became the largest copper producer in the world. To
do so, the pyrites had to be roasted, and the resultant sulfur dioxide was simply
released, causing devastation to the local countryside. An area miles around the
smelter became sterilized.

A price war ensued for three years, between Rio Tinto and TSCC, until the
two companies reached an "agreement," in 1876. Rio Tinto started shipping
pyrites to Britain in 1875, soon surpassing Tennant as the dominant supplier to the
largest sulfuric acid manufacturing nation in the world. By 1880, most of Europe's
sulfuric acid industry had switched to pyrites from sulfur, leaving the U.S. as the
only major importer of Sicilian sulfur for sulfuric acid manufacturing. After 1895,
the elemental sulfur market share in the U.S. also started to fall to pyrites (see
Table 2.2b). By the end of the 19th century, Rio Tinto was one of the largest mining companies in the world.

After the death of Matheson, J.J. Keswick led the company, followed by
Charles Fielding, who was chairman from 1905 to 1922. By World War I, the global sulfuric acid markets were now almost completely dependent upon pyrites,
especially from Spain. The pyrites market share soon found itself on a roller
coaster in the U.S. Trouble started during the unrestricted submarine campaign of
1917 which severely hampered shipments from overseas (both of Spanish pyrites
and Sicilian sulfur). At the start of the War, Spanish imports had been over one
million tonnes and had controlled 50% of the U.S. market; they had dropped
below 500,000 tonnes by 1918, even though acid demand was at record levels. In
July 1918, the U.S. government, through the War Industries Board, took control of
the production and allocation of pyrites and sulfur. As the TAC episode had led
the industry to switch to pyrites, World War I had the opposite effect on U.S.
industry, leading it to switch back to elemental sulfur, but now from the domestic
sources of the Frasch industry. By 1919, Spanish pyrites market share of the
American acid industry dropped to 21%; four years later, it was down to 13%.
Surprisingly, Spanish pyrites did not disappear all together, but continued to hold
a minority but steady market share. In 1937, the U.S. imported almost 500,000
tonnes (as sulfur) of pyrites from Spain.

Before World War I, more than half of the world's pyrites were supplied by
Rio Tinto, but this had dropped to 28% in 1935. Rio Tinto accounted for almost
60% of Spanish production, followed by TSCC. Between them, these two compa-

nies were producing more than two million tonnes per year. Following a distant third was Societe Francaise de Pyrites de Huelva. The situation was much different in Europe than the U.S., as pyrites remained the dominant source of sulfur for acid making. By the late 1930's, over 60% of the world's sulfuric acid was still being manufactured from pyrites, despite the changes in America.

These were turbulent times in Spain. During the Spanish Civil War and Fascist periods, the chairman (from 1924 to 1947) of Rio Tinto was Auckland Geddes (1879 - 1954), who had followed Lord Alfred Milner (1854 - 1925). In August 1936, the Rio Tinto area had fallen to Nationalist forces. Generally, business favored the Nationalist cause over the left-wing Republicans, including Rio Tinto who had been criticized in the British press for working with Franco. An infamous, callous statement by Geddes at the company's AGM in London was:

> *Since the mining region was occupied by General Franco's forces, there have been no further labour problems... Miners found guilty of troublemaking are court-martialled and shot.*

With the gaining of power by Franco and the fascists, Rio Tinto was in an increasingly difficult position. Most exports of the British-controlled company were already going to Nazi Germany. A marketing scheme was devised by the fascists to increase these shipments even more. A subversive joint venture, Hisma-Rowak, had been established between the German trading house Rowak and the Spanish firm Hisma (Sociedad Hispano-Marroqui de Transportes; partly controlled by the Franco government, and later 50% owned by Rowak). The slightly veiled purpose of Hisma-Rowak was to expedite raw materials for the Nazi war machine, including pyrites, from Spain to Germany. Rio Tinto officially, then, sold to a private Spanish firm, who in turn officially sold to a private German firm. Neither the Spanish or German governments were directly involved. Rowak, responsible for raw material imports into Germany, was controlled by a consortium of the leading German chemical companies, including I.G. Farben. Hisma had been set up in Spanish Morocco, by Johannes Bernhardt. The firm had the responsibility of establishing supply contracts for pyrites (and iron) from Spain and its colonies to Germany.

There was increasing pressure brought by the Franco government to control the business of Rio Tinto. Trouble started in late 1936, when two large parcels of pyrites were requisitioned by the government for Germany. During 1937, the pressure upon Rio Tinto continued. Hisma-Rowak, backed by the Franco government, forced Rio Tinto to increase their exports to Germany. Geddes was concerned about the aggressive moves of the government. The apprehension of the British executive of Rio Tinto was more practical than political in nature. The Franco government was forcing these sales at a discount of 6% below market value;

worse, profits of Rio Tinto now had to remain in Spain. While Germany may have been getting an edge on supply, British imports did not suffer; it was France who eventually lost their pyrites supply from Spain (Table 2.3).

Table 2.3. Pyrites – Spain; Exports in 1000 tonnes

Year	UK	France	Germany
1935	209	292	563
1936	201	327	464
1937	302	97	956
1938	320	1	896
1939	233	0	583

Leitz, C.: **Economic Relations Between Nazi Germany and Franco's Spain**, p. 75 (1996)

In early 1939, the simmering dispute between Rio Tinto and the Spanish government reached a head. Rio Tinto cut off Hisma-Rowak. On May 1st, the government countered by ordering Rio Tinto to cease all shipments outside of Germany and Italy. Franco started making plans to nationalize Rio Tinto; at the same time, Germany was demanding that they take over Rio Tinto (and TSCC) directly. While pressure from Britain eased Franco's acquisition ambitions, the Spanish government continued to exert indirect control over many of the company's activities. When the war started, Spanish exports were mainly limited to Axis allies. Between 1938 and 1944, production tumbled from 1.1 million tonnes to 0.4 million tonnes. By 1943, Geddes had determined to sell the problematic Spanish assets. Eleven years passed, before a deal was finally reached. Interference from the British and Spanish governments complicated the negotiations. In 1954, Rio Tinto sold two-thirds of its Spanish investments to a banking syndicate, and subsequently, the remainder was divested.

The Spanish Civil War and World War II finally ruined the Spanish pyrites industry. Shipments had been blocked during these years, and alternatives had been found. After World War II, many new sulfuric acid plants were constructed in Europe to replace those that had been destroyed, and U.S. expansion was bolstered by economic growth, especially by demand for phosphate fertilizers. These new plants all used elemental sulfur (Contact process). While Spanish pyrites production returned to pre-war levels by 1950 (see Figure 2.1 for the early history of production), their market share had seriously eroded as sulfur demand, overall, had more than doubled. Pyrites mining as a source of sulfur continued in Spain until 2002.

Before the discovery of the Frasch process, the only serious competitor to Sicilian sulfur was pyrites. During the pyrites golden years in the last quarter of

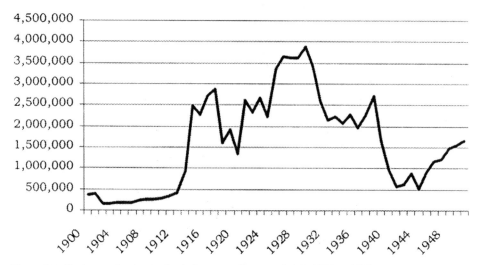

Figure 2.1. Pyrites production in Spain in tonnes (gross weight): 1900 to 1950.

the 19th century and the early years of the 20th, their market share of sulfur in all forms was even greater than Sicily. Pyrites, especially from Spain, were so successful that Rio Tinto along with ASSC and Union Sulfur, rightly hold the honor as being the world's first great sulfur companies. Some may legitimately argue that Rio Tinto should hold the title by themselves.

2.1.6.2 Italy

Another major producer of pyrites in Fascist Europe was Montecatini of Italy. In 1888, they began mining pyrites in Tuscany at Val Di Cecina, near the spa of Montecatini. Their major pyrites mines were Gavoranno, Ravi and Boccheggiano in Tuscany (650,000 tonnes per year); Agardo in Belluno (30,000 tonnes per year); and Brosso in Piedmont (35,000 tonnes per year). By 1910, Montecatini controlled all pyrites production in the country.

Italy, itself, was becoming a major sulfur consumer. The production of phosphate fertilizers had grown rapidly in Italy at the turn of the century. In 1898, sulfuric acid production was only 71,500 tonnes, but had surged to 650,000 tonnes before the beginning of World War I.

Between the wars, Montecatini flourished under the dynamic leadership of Guido Donegani (1877 - 1947), becoming the largest chemical company in Italy under his direction (1910 to 1945). Montecatini diversified into power stations, and purchased the two largest Italian producers of superphosphates, Unione Concimi and Colla e Concimi, in 1920. They expanded into explosives for their mining operations, becoming Italy's largest explosives company. During the 1930's, the Fascist government protected the company with tariffs, but, in return, Montecatini had to keep open certain unprofitable operations. Donegani had strongly supported the government. He became a Fascist member of parliament in 1921, and later became the president of the National Fascist Federation of Industries. Donegani, dressed in fascist uniform, was often seen with Mussolini. At the end of the war, on May 30, 1945, he was arrested by the Allies. After a brief investigation, he was released into the hands of Italian authorities in Milan. He had been kept at the San Vittore prison for over a month, when he was suddenly released on July 14th. The newspapers accused local authorities of being bribed by the rich industrialist. There were strikes at Montecatini plants in protest. The scandal soon passed. No charges were brought against the head of Montecatini. He spent his last days at Bordighera on the Italian Riviera, where he died in April 1947.

While the fall of Mussolini took Donegani with him, Montecatini continued to flourish after the war. After Donegani, Montecatini was run by Count Carlo Faina (1894 - ?), who had joined Montecatini in 1926. By 1948, Italy was again exporting pyrites; out of a production of 835,000 tonnes, 10% was exported. Between 1950 and 1955, total sales of the Italian conglomerate soared 85% to $276 million. By 1961, they were at $600 million. Montecatini was forcibly merged with Edison by the government to form Montedison four years later.

2.1.6.3 Norway

Behind Spain, Norway was the next largest producer of pyrites. The leading pyrites company in this country was ORKLA. Within the sulfur world, ORKLA became better known for its technology to convert pyrites into elemental sulfur, the ORKLA process. References to the process can be traced back to Agricola, who mentioned that sulfur was distilled from pyrites at Brambach and Harzgerode in Saxony, and Cromena in Bohemia during the 15th century.   Research by ORKLA had begun in 1919, and the company built a pilot unit at Lokken in 1927. Commercial production started at Thamshavn, Norway, in 1932 (closed in 1962), with production close to 50,000 tonnes in this year. At Huelva, Spain, Rio Tinto also adopted the ORKLA process (opened 1932; closed 1965). Their plant was the largest outside of Norway. When Rio Tinto attempted to sell sulfur into France, Sulexco (the U.S. export consortium) stepped in and made a deal with Rio Tinto to stay out of France. A third ORKLA process plant operated at San Domingos, Portugal (Mason and Barry; opened 1935; closed 1962); a small plant was also built in Hungary, and the process was believed to have been used in the U.S.S.R. Total sulfur production was over 90,000 tonnes in 1935 by the ORKLA process: 65,125 tonnes - Norway; 17,694 tonnes - Spain (1934 production); 8,566 tonnes - Portugal; increasing to over 140,000 tonnes of sulfur per year during World War II: Thamshavn at more than 100,000 tonnes; Huelva at 27,000 tonnes and San Domingos at 15,000 tonnes. The plants had all closed by the mid-1960's. Other processes using pyrites (and other smelter gas) to produce sulfur were developed:

Sweden at Ronnskar – 1930's
Canada - Cominco at Trail, BC – 1935 to 1943
U.S. at Garfield, UT – 1940
Canada – Noranda at Port Robinson, ON – 1954 to 1959
Finland – the Outokumpu (ICI) process at Kokkola – 1962 to 1977
Canada – INCO at Port Colborne, ON – 1957
Canada – INCO at Thompson, MB
Canada – Allied Chemical at Falconbridge, ON – 1972 to 1973

2.1.6.4 Canada

Related to pyrites being converted to sulfur was the utilization of smelter off-gases. In 1935, Cominco opened a plant at Trail, BC, to convert sulfur dioxide off-gas into elemental sulfur, marking the first production of the yellow element in Canada. Sulfur dioxide was passed through a coke bed in the presence of oxygen, producing carbon monoxide and elemental sulfur. The plant operated until July 1943, when converted to manufacture only sulfuric acid. At the time, Cominco was the major source of exports of elemental sulfur to the U.S., at 27,750 tonnes. This volume was trivial compared to U.S. production, but few other countries were producing any elemental sulfur. Canada was about the fifth largest producer in the world at the time.

In 1954, elemental sulfur was also produced in Canada from pyrites, when Noranda produced sulfur (18,000 tonnes per year) at Port Robinson, ON (near Welland). The plant also supplied sulfuric acid (from pyrites) to the nearby phosphate fertilizer facility of Cyanamid (Welland Chemical Works). This $4.7-million plant at Port Robinson greatly increased the demand for pyrites until the plant closed in 1959. In 1957, INCO began producing elemental sulfur from nickel smelting at Port Colborne, ON, and, later, at Thompson, MB. The last smelter-source elemental sulfur in Canada came from Allied Chemical; they started producing sulfur from pyrites at Falconbridge, ON, in early 1972, but the facility closed a year later.

Pyrites mining in this country had started with the early explorers looking for the yellow element. However, this element was not sulfur but gold! After his third voyage to Canada in 1541, Jacques Cartier (1491 - 1557) brought back to Paris, allegedly, diamonds and gold. The deceptive treasures turned out to be quartz and pyrites ("fool's gold"). The latter material had also duped Martin Frobisher (1535 - 1594). In 1576, he had found some black earth in the frozen reaches of the Canadian North that allegedly contained gold. His second voyage the following year was undertaken solely to collect the "valuable" ore; he brought back on this arduous journey 200 tonnes of the worthless material to England. A third voyage of Frobisher's follies was made in 1578 to *Meta Incognita* (part of Baffin Island) as the northern land was proclaimed. The intent of the latest voyage was to form a colony to exploit the rich resources. While the colony never happened, he did return with more ore. By this time, the "gold" had been determined to be worthless pyrites. Frobisher's financial backers were flabbergasted, and his reputation was ruined by the incident. This scandal put an end to the short-lived and unintentional pyrites operations in the New World. Pyrites mining in Canada would not begin again until 1866 at Capelton, Quebec.

The American General Adams operated a mine at Capelton, near Sherbrooke, from 1866 to 1871. The Adams' operation was the first pyrites mining in this country since the days of Frobisher. Capelton and its pyrites were the centre of the early Canadian sulfur industry:

Canadian Copper Pyrites and Chemical Company – 1870's
Nichols Chemical Company – 1880 (closed in 1939)
Oxford Copper and Sulfur Company – 1880's
Grasselli Chemical Company – 1887.

Production of pyrites in Canada was over 15,000 tonnes (based on sulfur content of the pyrites) in 1886, the first year that data was collected. Production was between 10,000 and 25,000 tonnes per year until the end of the century. Quebec generally remained the major source of pyrites in Canada until 1900. The earliest pyrites operation in Ontario was near Brockville. After 1900, several deposits were developed in Ontario, including Vermilion Mining, near Sioux Lookout; Northland Pyrites near Timagami; and Goudreau Lake. Pyrites mining in Ontario had ceased by the mid-1920's. In 1916, pyrites production as a source of sulfur began in B.C.

A fascinating little company in Canada was producing pyrites to recover sulfur from "coal gas." In 1882, the Vesey Chemical Company of Montreal commenced burning "spent oxide" from the city's gasworks into sulfur dioxide. The spent oxide was iron ore that had been used to remove hydrogen sulfide from coal gas. The plant was purchased by Nichols Chemical in 1902 and shut down.

Before World War II, the Canadian "sulfur" industry was still basically only pyrites. Not on the scale of Spain, Canada was still a significant exporter of this material into the U.S. During World War I, over one hundred thousand tonnes of pyrites for sulfuric acid manufacture were produced in Canada. The surge in demand was not only led by increasing acid demand, but more importantly the disruption in shipments of pyrites from Spain into North America. After the war, pyrites production in Canada plummeted because of pressure from the emerging Frasch production in the U.S. Discretionary pyrites production in Canada had ended by 1939. Pyrites were still produced, but only as a by-product of the flotation of metal-sulfide ores. The major by-product producers of pyrites were Noranda in Eastern Canada and Britannia Mining and Smelting (liquidated in 1959) in Western Canada. Pyrites production peaked in 1947 at almost 500,000 tonnes, but then went into a decline. In the 1951 issue of the **Canadian Minerals Yearbook**, there was an interesting change in the report. Instead of being under "Pyrites and Sulfur", as in previous issues, the title was changed to "Sulfur and Pyrites." By the 1959 issue, the word "Pyrites" was no longer in the title of the section on sulfur, reflecting the changing industry. Much of the later pyrites demand was internal, as the pyrites producers themselves converted their product into sulfuric acid. Otherwise, pyrites were used only at a few locations near deposits where they had a delivered cost advantage over Frasch sulfur. In 1973, Noranda stopped pyrites sales as a source of sulfur, as their major customers in the NEUS had switched to elemental sulfur. The last pyrites production in Canada ended in 1982. In 1987, the last pyrites mine closed in the U.S.

2.1.6.5 China

Today China is the only remaining major pyrites producer. At the same time, this country is also the largest importer of elemental sulfur in the world. Pyrites mines include the Hsiangshan mine at Anhwie, the Ying-te mine at Kwangtung, and a mine at Yunfu. Today, China produces more than three million tonnes of pyrites, as sulfur, each year.

2.2 SULFURIC ACID MARKETS

2.2.1 THE LEBLANC PROCESS

The Chamber process revolutionized the sulfur world. Until then, acid was a novelty product which could only be used in miniscule quantities because of cost. With the developments of Roebuck and others, sulfuric acid was about to become the most widely used chemical in the world!

The major price reduction of sulfuric acid initiated by the Chamber process led to new studies on this versatile acid. One of the first new developments, taking place in the 1750's, was for textile bleaching. The initial work was done by Francis Home (1719 - 1813). The technology was quickly adopted by the end of the century. One of the major early users of this technology in Britain was Richard Bealey & Company, at the beginning of the 1790's. The firm was later run by Richard Bealey (1810 - 1896).

Another new technology led to explosive growth, and not because of gunpowder, for sulfuric acid. While manufacturing processes for mineral acids had been known for centuries, their antipodes in the chemical world, the alkalis, were limited, a major source being burnt-wood ashes and other plant residues. In the last quarter of the 18th century, a growing demand for these products had emerged for the soap industry in Britain and the glass industry in France. Soda ash (sodium carbonate) was identified as the alkali with the greatest scientific potential for a

new manufacturing route to meet the needs of these bubbling industries. Ironically, sulfuric acid was the key to its chemical manufacture.

The fledgling science of chemistry was still very crude. The discovery of
alkali was one of the first industrial processes discovered not by pure luck, but by
a deliberate attempt to utilize the chemical knowledge, weak as it was, of the day.
In 1737, Duhamel du Monceau (1700 - 1782) had found that soda ash and salt,
were both sodium derivatives. With salt being so readily available, it would be an
obvious choice to try to convert to soda ash; a suggestion that Monceau himself
had made. With so little information to base their research upon, the chemical
quest, one of the first since the alchemists had tried to produce gold, would last for
half a century. A legion of the leading scientists was drawn to this problem. While
many scientists contributed to the step-wise development of the technology, the
real honor, albeit somewhat unfairly, is available only to the one who finds the last
piece of the chemical puzzle. However, their research has been distilled from the
test tubes of others who preceded them.

In terms of the history of the sulfur industry, the most important development
took place in 1772, when the famous Swedish chemist Carl Scheele (1742 - 1786)
reacted salt with sulfuric acid to produce sodium sulfate. The sulfate was a more
versatile reagent than the chloride, increasing the chances of producing alkali.
After the work of Scheele, focus fell on the conversion of the sulfate into soda ash.
Reactions were soon discovered to carry out this process, but none worked well
enough to be commercially viable. In Britain, alkali research was carried out by
James Keir (1735 - 1820), a friend of James Watt and Joseph Priestley, in the early
1770's. He produced sodium sulfate from salt and then converted it to sodium
hydroxide (an alkali) with slaked lime. He opened a small plant with Alexander
Blair, at the Tipton Chemical Works in Bloomsmithy, in 1780. Keir cleaned up,
after the site became one of the largest soap manufacturers in Britain. Another
small facility was operated by Alexander Fordyce (patent in 1781) and his brother
George at South Shields. The discoverer of the Chamber process, John Roebuck
with James Watt, attempted to produce alkali without success. Other early British
studies in the field were carried out by Brian Higgins, Richard Shannon (patent in
1779) and James King (patent in 1780). In Germany, Andreas Sigismund Marg-
graff (1709 - 1782) was working in this area. French scientists, especially, were
leading this field, including Malherbe (1733 - 1827) in 1777, P-L. Athenas, Jean-
Claude de la Metherie (1743 - 1817), Hollenweger, Jean Antoine Chaptal (1756 -
1832) and Jean Henri Hassenfratz (1755 - 1827). An alkali plant had been build
by Louis-Bernard Guyton de Morveau (1737 - 1816) and Jean Antoine Carny at
Croisac (Brittany) in 1782, but the venture failed.

While a few plants were built in Europe, none of them survived for very
long. These pioneering researchers had produced soda ash from sodium sulfate, in

most cases based around the use of coal and iron. While these developments were chemically interesting, they were uneconomic or produced low quality material. A better technology was still required.

The discoverer of the right process was Nicolas Leblanc (1742 - 1806) in 1789 (his patent was issued on September 25, 1791). Much of the process had been discovered by those mentioned above. The first step, the reaction of sulfuric acid with salt to form salt cake (sodium sulfate), had been reported by Scheele. The second step included coal, as had been used by Malherbe, and others. Leblanc's contribution was the addition of chalk (or limestone) to the second reaction, thus producing a practical invention:

$$H_2SO_4 + 2NaCl \text{ (salt)} \rightarrow Na_2SO_4 \text{ (salt cake)} + 2HCl$$

$$Na_2SO_4 + CaCO_3 \text{ (chalk)} + 2C \text{ (coal)} \rightarrow Na_2CO_3 \text{ (soda ash)} + CaS + 2CO_2$$

Leblanc had stumbled across the simple, but critical, addition of chalk to make the reaction work better. His inspiration, whether by the legendary luck of the great chemists, a suggestion by de la Metherie or Dize, or scientific skill, is unknown and from a practical sense does not matter. The simplicity of the modification strengthens, not weakens, the importance of the discovery. An industrial process has more chance of surviving with basic ingredients than exotic ones, and the Leblanc process was about as simple as it got. While apparently straightforward, the facts spoke for themselves; no one else had discovered it even though the search for the soda ash route was one of the most congested research areas of the day.

The discovery was only part of the importance of Leblanc's contribution. He also commercialized the technology. Leblanc built his leading-edge technology facility at the "La Franciade" works in St. Denis, outside of Paris in 1791. Partners in the venture were Louis Philip II, the Duke of Orleans (1743 - 1793), Michel Jean Jerome Dize (1764 - 1852) and Henri Shee. The small facility, measuring 10 x 15 meters, produced 0.25 tonnes per day of soda ash. Unfortunately for the inventor and his partners, France was soon in the turmoil of the Revolution, and the major investor was on the wrong side. By the order of the Committee of Public Safety, the facility was confiscated by the state in 1794, not because of Leblanc, but because its major investor had been the Duke of Orleans, who had been sent to the guillotine the year before. The loss of his plant was one thing, but Leblanc also lost something even more valuable, his technology. The patented process was publicly released by the Committee in 1797, in the French chemical journal *Annales de Chimie*, for anyone to use royalty free.

The decrepit plant was returned to Leblanc in April 1801 by Napoleon, but Leblanc did not have the funds to retrofit the abandoned site. His patent protection was never returned. A broken man, he committed suicide in 1806. Leblanc has been labeled a "martyr of technology." However, even in death, the fall of Leblanc was not yet over. Four years later, an attempt was made to even take his discovery away from him. In a technology court-martial, Dize claimed that he had been the one to suggest the addition of chalk, the essence of the invention. The controversy grew in the following decades. Finally, in 1856, the French Academy put an end to it by officially declaring that the invention belonged to Leblanc alone. Even so, an article in 1907, by H. Schelenz (*Chem.-Ztg.*, **30**, p. 1191), advocated that the 25-year-old Dize was the true discoverer of the process.

The actions of the Revolutionary government had allowed others to freely benefit from the discovery of Leblanc. In 1806, a Leblanc plant was opened by Payen at Dieuze, and others opened in Rouen and Lille. In 1810, the first Leblanc plants were operating in Marseilles, which became the heart of the French soda ash industry. By the end of the decade, total alkali production surpassed 10,000 tonnes per year in France. Sulfuric acid demand surged with it. In 1815, France was producing 20,000 tonnes per year of sulfuric acid. In 1822, a major Leblanc plant was opened in Chauny. Chemical production on this scale had never been seen before.

The first plant in Germany was not built until 1840 by von Hermann & Sohn in Schonebeck; other early works were opened by the Italian chemists Jean Angelo Grasselli, born in Torno, Italy and father of Eugene Grasselli (see above), at Mannheim, and the Giulini Brothers at Wohlgelegen. However, the first major plant in Germany was not built until 1851 in Heilbronn. Later, in 1866, BASF built a Leblanc plant at their Ludwigshafen complex. The technology was never employed in North America.

During the first quarter of the 19th century, France dominated the alkali world, but the British were poised to soon surpass them. So dominant did the latter become that the Leblanc process became more British than French! However, the Leblanc process had been delayed being implemented in this country, because a salt tax inhibited its economics. The salt tax was one of the more controversial aspects of the consumption tax system that had been implemented to diversify away from solely a property tax system. Introduced in 1644, it had been briefly repealed but the government of Robert Walpole (1676 - 1745) reinstated the levy in 1732. This hefty tax of £33 per tonne was an economic deterrent for any industrial application of salt, especially considering that production costs of the Leblanc process were otherwise only £1 per tonne. Only two small Leblanc plants had been built in Britain; one by William Losh, who had earlier worked with Archibald Cochrane (1748 - 1831; Earl of Dundonald) on alkali manufacturing and the tar industry, at the Walker Collier in Tyneside in 1814; and the other in Scotland at the celebrated St. Rollox Works of Charles Tennant. Losh had learned about the technology from his friend, the revolutionary Jean-Paul Marat (1743 - 1793), and in 1802, he had visited Leblanc's closed plant at Saint-Denis and made sketches of the process. The Tennant plant was small; the dimensions of the associated Chamber acid reactor were 12 feet by 10 feet. James Muspratt may have also operated a small Leblanc plant in Dublin. The salt duties were removed between 1818 and 1825. Once this tax was lifted, alkali plants proliferated across the nation.

The Leblanc plants were the origins of the giant British chemical industry of the 19th century, being the predecessors of the massive chemical complexes that were built by the Mersey and Tyne rivers and in Glasgow. The first of the post-salt tax facilities was installed by the legendary chemical entrepreneur, James Muspratt (1793 - 1886) in 1823, near Liverpool, followed by another plant five years later at St. Helens. The size of the Chamber plant at Liverpool (112 feet x 24 feet) reflects the change in scope of these facilities. Muspratt's business partner at St. Helens was Josias Christopher Gamble (1776 - 1848); whose nephew James Gamble (1803 - 1891), along with his brother-in-law William Proctor, formed Proctor & Gamble in the U.S., in 1837. Jos. C. Gamble and Company had been producing sulfuric acid in Dublin from 1814 to 1821.

The fundamental flaw of the Leblanc process reared its ugly head. The new process had serious environmental effects, as the waste hydrochloric acid fumes devastated the local countryside. As early as October 5, 1827, an article appeared in the *Liverpool Mercury* protesting the "sulfureous smoke." Already innocent sulfur was being maligned. The emissions did include some sulfur dioxide from the manufacture of sulfuric acid, but the more serious issue was the gaseous hydrogen chloride. Muspratt was flooded with law suites. His alkali operations were shut down, except at Newton, which had opened in 1830. Here, to diffuse the gas, a 400-foot stack was constructed. Finally, he was forced to close this plant as well. New operations were constructed at Widnes and Flint in 1850. The government passed the Alkali Act of 1863, forcing the Leblanc producers to capture (most) of their by-product hydrochloric acid.

The new technology and ready market did not mean instant commercial success. The British soap manufacturers were reluctant to use the new product in their operations. Muspratt had to first give the product away for them to try it. Afterwards, though, demand soared. One is left with the impression that anyone could open a successful chemical business during these boom times; this was not the case. Later, others set up operations in the heart of the British soap industry and failed; for example, Edward Rawlinson, William and Samuel Clough, and William Kurtz (1782 - 1846). Generally, these operations were either too small and/or under funded to succeed.

The British chemical industry was founded by the Irish (Muspratt) and the Scots. The latter was Charles Tennant, who built a major Leblanc plant in Scotland at St. Rollux, in 1825. Following his death in 1838, his son John Tennant (1796 - 1878) became head of the company. During his tenure, a giant 435-foot stack was built at the St. Rollox Works in 1842, surpassing the stack at Newton. This Glasgow landmark became known as "Tennant's Stalk." The Stalk was the largest industrial stack in the world. The Stalk stack was dismantled during the 1920's, and St. Rollux was closed in 1964. The company still operates today.

British alkali capacity had surged past their French competition. In 1852, French production, mainly in Marseilles, was 45,000 tonnes, while British was 140,000 tonnes. The growth of this industry was mind-boggling. The "basic" economy of the country was dominated by acid! By 1865, there were eighty-three Leblanc plants operating in Britain. As the last quarter of the 19th century approached, there seemed to be no end to the growth of the Leblanc process and its sulfuric acid demand. However, this was but the calm before the storm. The industry would be on its last legs by the end of the century, much to the chagrin of the sulfur industry and, especially, the Leblanc producers themselves. The French technology in Britain, and elsewhere, would fall to a Belgian development. The Leblanc companies were broadsided by a more efficient technology to produce alkali known as the Solvay process. The major difference between the two alkali technologies was that the intermediate sodium sulfate was replaced by sodium bicarbonate and, more importantly to the history of sulfur, sulfuric acid was no longer required. There were also almost no gaseous emissions with the new process. As with the Leblanc process, the Solvay technology had long been under development. The initial reaction had been discovered by Augustin Jean Fresnel (1788 - 1827) in 1811, and in 1838, John Hemming and Harrison Dyer were issued a patent on the process. However, they and others, including Muspratt, could not get it to work on a commercial scale. Ernest Solvay (1838 - 1922) of Belgium had improved the older process to manufacture soda ash from salt and ammonium carbonate in 1861, and had opened his first plant in Couillet, Belgium, four years later. Solvay licensed his technology to his friend Ludwig Mond (1839 - 1909) for development in Britain in 1873. Mond, himself, had been involved in the Leblanc industry of Germany. Mond with John Brunner (1843 - 1919) introduced the Solvay technology to Britain, building a plant at Northwich. The Solvay invasion had begun. In 1874, only 4,000 tonnes of soda ash had been produced in Britain by the new technology, while the Leblanc process was the source of almost 200,000 tonnes. From this small beginning, the process of industrial evolution had started, and there was no stopping it. Among the staggering number of alkali works in Britain, the shift towards the Solvay process came relatively quickly. The total number of alkali works in Britain was more than 1,000 in 1888:

866 in England, 131 in Scotland and 44 in Ireland. This industry was consuming over 725,000 tonnes of salt (544,000 tonnes - Leblanc; 181,000 tonnes - Solvay). Worldwide, the Leblanc production had leveled off by 1880, and then started to decline, dropping to 200,000 tonnes by 1900, as Solvay soda ash displaced the product of the older technology (see Table 2.4).

Table 2.4. Soda ash production in 1000 tonnes by Leblanc technology

Year	Britain	World
1850		150
1852	140	
1863		300
1865		374
1870		447
1875		495
1878	179	
1800	241	545
1882	212	
1884	185	
1885		435
1886	151	
1890		390
1895		265
1900		200
1902		150
1905		150
1911		130
1913		50

Source: Hou, T-P.: **Manufacture of Soda**, 1933; Haber, L.F., **The Chemical Industry of the 19th Century**, 1958.

An interesting side-note was that the Leblanc industry in Britain almost never happened (to any degree). The basic process that was improved upon by Solvay had been known for many years. James Muspratt had tried the technology in 1838 but could not get it to work. The process would have been of special interest to Muspratt, since he was already being harassed by the discharges of the Leblanc technology. If he had been successful, the "Solvay process" would have taken over the British chemical industry almost half a century earlier. The Leblanc industry, then, would have been relegated to a footnote in the industrial history of the chemical industry. The impact on the sulfuric acid and sulfur industries would have been unimaginable.

2.2.2 PHOSPHATE FERTILIZERS

While the Leblanc process had shot sulfur into the stratosphere, another development would catapult the yellow element to levels seen by no other chemical. So great was this new market that the demise of the Leblanc process was barely noticed by the sulfur industry. Well over a century later, fertilizers are still the dominant market for sulfuric acid.

In 1840, Justus von Liebig had made his landmark discovery of the critical importance of nitrogen-based fertilizers in agriculture and had recommended the use of phosphates. An attempt to manufacture phosphate fertilizers had been made by James Sheridan Muspratt (1821 - 1871), the eldest son of James, in 1843. He had been a student of Liebig, and using the latter's technology, he opened an early fertilizer plant at his father's facility at Newton using the Liebig process. The project failed, for Liebig had made a mistake by promoting the use of a water-insoluble fertilizer.

The idea of artificial fertilizers had long been known, but Liebig had placed this study on a firm scientific basis. Work in this field had been carried out by James Murray (1788 - 1881), who as early as 1817 had carried out field trials of bones dissolved in sulfuric acid. In 1842, his son John Fischer Murray filed a patent to manufacture a fertilizer from calcium phosphate and sulfuric acid.

Credit for the industrial success of phosphate fertilizers was not given to Murray, but to John Bennett Lawes (1814 - 1900), who also filed a patent process for manufacturing superphosphate fertilizers in 1842 (Brit. Patent 9353). Lawes was better at commercializing the technology than Murray and bought the rights to the Murray patent in 1846. His product was dubbed "artificial manure" or just "manure." In the traditional wet process for manufacturing phosphate fertilizer, phosphate rock is treated with 93% sulfuric acid for eight hours, producing single superphosphate (fertilizer); the original sulfur in the sulfuric acid ends up as waste calcium sulfate (gypsum):

$$2H_2SO_4 + Ca_3(PO_4)_2 \text{ (phosphate rock)} \rightarrow Ca(H_2PO_4)_2 \text{ (single superphosphate)} + 2CaSO_4$$

Lawes built his first plant at Deptford in 1844. Already by 1855, 60,000 tonnes of phosphate fertilizers were being applied in England alone, most of the product coming from Lawes, and at the end of the decade, a dozen phosphate plants were in operation, mainly in Britain and Germany. In 1906, a superphosphate plant was installed in Spain by Rio Tinto. The rapid development of the fertilizer industry, especially phosphate fertilizers, is mind-boggling.

The technology spread to North America. In 1869, the first phosphate fertilizer in Canada was produced by the Brockville Chemical and Superphosphate Company, owned by Cowan and Robertson. The Brockville Chemical and Superphosphate company operated between 1869 and 1884. The plant closed because the local pyrites source for sulfuric acid manufacturing became exhausted. Three years later, the owners opened a new superphosphate and sulfuric acid plant in Smiths Falls, ON (closed in the 1920's); the sulfur for the Standard Fertilizer and Chemical Company plant came from Sicily and Japan. Other plants opened in Canada during the 1880's:

      1886 - W.A. Freeman Fertilizer Company in Tweed, ON (closed 1911)

      1889 - Nichols Chemical Company opened the Capelton Chemical and Fertilizer Works, near Sherbrooke, QC (closed in 1902)

      1889 - Provincial Chemical Fertilizer Company, in Saint John, NB

A related fertilizer was ammonium sulfate. In 1901, Dominion Steel and Coal Corporation opened a sulfuric acid plant (Chamber process) to produce ammonium sulfate in Sydney, NS. The plant first used pyrites from Newfoundland, but later switched to imported elemental sulfur. The ammonia was a byproduct from their coke ovens. This facility was the first ammonium sulfate plant in Canada. By 1946, 400,000 tonnes of sulfuric acid (100% basis) was used by the fertilizer industry, representing over 75% of the acid consumed in this country. Canada, though, never became a major fertilizer producer.

The first American superphosphate plant opened in 1850 by William Davison and T.S. Chapell of Baltimore. Other early producers in this decade were Potts & Klett and Moro Phillips of Philadelphia. The introduction of the phosphate fertilizer industry created the first significant sulfuric acid market in the U.S. This country would eventually dominate phosphate fertilizer production, thanks mainly to massive phosphate deposits in Florida, which were first mined in 1888. Now, in the early 21st century, the U.S. market for sulfuric acid in the phosphate fertilizer industry is over 25 million tonnes, representing over 8.0 million tonnes per year of sulfur. The largest producer of phosphate fertilizer in the world is Mosaic, formed from the merger of IMC Global and Cargill Crop Nutrition in October 2004 (see Table 2.5).

Table 2.5. Phosphates; US mines in 2004

Company	Location
Mosaic	Fort Green, FL
	Four Corners, FL
	Hookers Prairie, FL
	Hopewell, FL
	Kingsford, FL
Mosaic	South Fort Meade, FL
	Wingate Creek, FL
PCS	Hamilton, FL
	Aurora, NC
J.R. Simplot	Caribou, ID
	Uintah, UT
CF Industries	South Pasture, FL
Agrium (Nu-West Industries)	Rasmussen Ridge, ID
P4 Production (Monsanto)	Enoch Valley, ID

While the U.S. has long been the largest phosphate producer in the world, two other regions have become major global suppliers: Morocco and China. For most of the 20th century, Morocco (and Tunisia) has been the largest producer of phosphate rock after the U.S. (see Table 2.6). Morocco contains the largest phosphate rock reserves in the world, with a major deposit at Khouribga. Morocco changed from simple mining to becoming a major phosphate fertilizer producer, with facilities at Safi and Jorf Lasfar. Imports of sulfur into the country are controlled by Office Cherificn des Phosphate (OCP). In November 1997, a liquid handling terminal was installed at Jorf Lasfar by OCP. There are 300,000 tonnes of liquid storage which supplies the phosphate facilities by a 1.5-km pipeline. Liquid sulfur was being supplied by MG Chemiehandel of Germany and Elf Aquitaine (now Total) of France. As of the beginning of the 21st century, Moroccan sulfur demand is more than two million tonnes per year.

Table 2.6. Phosphate rock mining in the world in 2004

Country	P_2O_5 content, MM tonnes
US	10.4
Morocco	8.5
China	7.7

Table 2.6. Phosphate rock mining in the world in 2004

Country	P_2O_5 content, MM tonnes
Russia	4.0
Tunisia	2.4

Source: **U.S.G.S. Minerals Yearbook**

China has become the fastest growing phosphate producer in the world. Between 1986 and 1996, sulfuric acid production in this country doubled. Demand for sulfur escalated even more as the government promoted self-sufficiency in the agriculture sector, with the issuing of the Ninth Five-Year Plan in 1995. A massive growth in phosphate fertilizer capacity took place, and with it, total sulfur imports skyrocketed: 0.8 million tonnes - 1997; 1.3 million tonnes - 1998; 2.0 million tonnes - 1999; 5.3 million tonnes - 2003, 6.8 million tonnes - 2004, 8.0 million tonnes - 2005. This enormous market growth has strained the global sulfur supply network. Canada is the dominate sulfur trading partner with China. There has been a long history of sulfur trade between the two countries, beginning in 1972, with a shipment of just under 200,000 tonnes. In 1986, China suddenly stopped importing sulfur from anyone. The government had established a program whereby China would become self-sufficient in sulfur, by developing new pyrites deposits. Despite becoming the largest pyrites producer in the world, China had to re-enter the import market in 1989. Canadian exports to China were around 100,000 tonnes per year in the early 1990's. As fertilizer expansions took off in the mid-1990's, Canada established itself as the major overseas trading partner with China for sulfur:

1996	0.7 million tonnes
1997	0.2 million tonnes
1998	0.4 million tonnes
1999	1.3 million tonnes
2000	1.7 million tonnes
2002	2.2 million tonnes
2003	2.3 million tonnes
2004	3.7 million tonnes

Since the beginning of the new millennium, China has been the largest importing country of sulfur in the world, mainly for phosphate fertilizer manufacture.

The Leblanc process was the beginning of the modern chemical industry. What followed was the first global chemical market. Sulfuric acid, more so than the alkali itself, was the yellow heart of this worldly enterprise. During the 20th century, the demand for sulfur (in all forms) has increased from 1.4 million tonnes in 1901 to 59.3 million tonnes in 2000, and the dominant market has remained the production of sulfuric acid, especially for the phosphate industry. This material was the quintessence of the Industrial Age, or, more correctly, the Sulfur Age, and continues to be so today.

Chapter 3

NATIVE SULFUR — SICILY

First the Leblanc process for alkali manufacturing, followed by the phenomenal growth for phosphate fertilizers, created an insatiable industrial thirst for sulfur around the world. Until the early 20th century, the only source of elemental sulfur that could supply the burgeoning demand was Sicily.

3.1 ANCIENT SULFUR

While the Sicilian sulfur industry is often mentioned as thriving during Roman times, or even before, the ancient literature makes no mention of this. The relevant references where such documentation would have been expected to be found are the works of Strabo, who summarized the geography of the Roman world; Diodorus Siculus, who wrote a comprehensive history of the world; and Pliny, the prolific compiler of global knowledge.

Strabo (63 B.C. - 24 A.D.) writes about mines within the Roman world in his **Geography**. Among the specific types of mines mentioned were iron, copper, arsenic, gold, silver, lead, tin and emerald, but not a word about sulfur mines. He goes on to report that all sorts of mines existed in Italy (6.4.1), and of ancient copper mines at Temesa, in south-western Italy, near Sicily (6.1.5; 12.3.23). No mining in Sicily itself was reported.

If any classical writer would have mentioned the sulfur mines of Sicily, it would have been expected to be Diodorus Siculus (fl. 50 B.C.), who was from Sicily. He mentions the mining of tin in Britain, tin and silver mining in Spain, iron mines in Elba, and gold mines in Egypt, but nothing about sulfur, even in his home country.

Whereas sulfur mining was ignored by most ancient scholars, Pliny describes in detail this ancient industry. Sulfur was mined in Italy, at Naples, Campania, and the Aeolian Islands (35.50.174). Pliny comes close to Sicily with the sulfur mines of the nearby Aeolian Islands, but nothing on the island, itself, is mentioned. He goes on to report on various geological aspects of Sicily, including the importance of bitumen (but not sulfur) at Girgenti (35.51.179), which lays in the heartland of the later sulfur industry in Sicily!

Where's the sulfur? Italy was a major source for the Empire's limited sulfur needs, but the island of Sicily, specifically, was not one of them. Pliny and the other ancient scholars clearly imply that no major sulfur mines existed on Sicily at the time of the early Roman Empire.

3.2 RENAISSANCE SULFUR

The growing munitions trade sparked technological reviews on sulfur mining and processing:

- Vannoccio Biringuccio (1480 - 1539) prepared one of the first works on the sulfur industry in 1540. His book, **Pirotechnia**, was published after the death of its author. The Italian metallurgist's chapter on sulfur is found in Book II. To purify crude sulfur, the ore was placed in pottery vessels containing a spout near the top. The vessels were heated and the sulfur distilled into a receiving vessel. Among the uses of sulfur of the day were for medicine, bleaching wool, silk and women's hair. The largest market was gunpowder. Sulfur was found across Italy, including the Aeolian Islands and Mount Etna, one of the earliest references to sulfur in Sicily.
- Georgius Agricola (1494 - 1555) wrote the geology classic **De Re Metallica (On the Nature of Metals)**. This practical work was about the business and industry of mining. Agricola had read the book of Biringuccio and had studied in Northern Italy from 1524 to 1526. His section on sulfur (p. 578 to 581) describes its purification in the same manner as Biringuccio, but in greater detail. Agricola, though, makes no mention of Sicily.
- Luigi Ferdinando Marsili (1658 - 1730) visited the sulfur mines of Romagna, in Northern Italy, in 1675. In the 1720's, he prepared maps and illustrations of the industry.
- Vincenzo Masini, in 1759, wrote the most unusual review of the Italian sulfur industry. The uniqueness did not originate from its contents but its style; it was a poem, called **Lo Zolfo (Sulfur)**. The work was dedicated to the legate of Romagna.

The Cesenate/Romagna region of Italy was one of the major sulfur producing regions in Europe by the 16th century. Sulfur ore was distilled into sulfur at Cesena, Volterra, and Pozzuolo. A local sulfur monopoly in the Cesenate territory (part of the Papal States) was given to Bartolomeo Valori of Florence, by Pope Clement VII (1478 - 1534). The region had become a major trading center to the lucrative munitions industry. A critical logistics port was Cesenatico. The government under Cesare Borgia (1475 - 1507) hired Leonardo da Vinci (1452 - 1519) to design the expansion of the harbor in 1502.

The works of Renaissance scholars continue to record the importance of the mainland sulfur mines of Italy, but, again, Sicily is seldom mentioned. Even with Biringuccio, the Sicilian sulfur that he mentions is solfataric, not the anhydrite-gypsum deposits that made the island the sulfur capital of the world.

3.3 INDUSTRIAL REVOLUTION ("SULFUR AGE")

3.3.1 SICILY

Despite the fact that the early literature makes little mention of sulfur mining in Sicily, this region would emerge as the dominate global producing nation of the Industrial Age. A massive sulfur vein, called "Sicilian gold," extended over one hundred miles from Mount Etna to Girgenti (Agrigento). The production centre was at Caltanisetta (pop. = 16,000 in 1850; 43,000 in 1901), with major mines in neighboring Cattolica, Licata, Caltascibetta, Centorbi, and Sommatino. The commercial development of this massive sulfur source was likely not developed until the early 18th century. The industry was already established when the famous Floristella-Grottacalda mine opened in 1750.

The major market for Sicilian sulfur was Marseilles, where sulfur refineries and sulfuric acid plants were located. Even before the introduction of the Leblanc process, sulfur imports from Sicily into Marseilles were about 2,000 tonnes per year (see Table 3.1). French imports of sulfur were disrupted by the French Revolution and the subsequent reign of Napoleon, as war spread across Europe. With the overthrow of Napoleon and peace returning to Europe, France, where most of the early Leblanc processes had been installed, was the largest customer. Marseilles was the leading chemical manufacturing center in the world in the early 19th century. In 1815, 6,500 tonnes of sulfur were produced in Sicily, most of which went through Marseilles. By the middle of the 19th century, sulfur had become one of the top exports from Sicily:

- 1839 (in million French francs; total exports were 34 million FF) – sumac spice (6.6); dried fruit (4.0); wine (4.0); sulfur (3.4); citrus fruit (3.0); and olive oil (2.4);
- 1852 (in British pounds): sulfur – £392,000; citrus fruit – £362,000; olive oil – £163,000; wine – £106,000.
-

Table 3.1a. Sulfur – Marseille.

Year	Sicily imports, tonnes	Value, francs
1784 to 1789		
1784	100	n.a.
1785	2,525	n.a.
1786	1,808	n.a.
1787	1,144	246,000
1788	2,800	491,500
1789	n.a.	246,000

Table 3.1a. Sulfur – Marseille.

Year	Sicily imports, tonnes	Value, francs
1832 to 1841		
Year	Sicily imports, tonnes	Total imports, tonnes
1832	12,382	12,507
1833	12,036	12,082
1834	17,331	n.a.
1835	18,106	n.a.
1836	24,953	n.a.
1837	17,535	n.a.
1838	36,616	n.a.
1839	12,513	13,013
1840	16,649	17,250
1841	n.a.	9,321

Source: Jules Juliana, **Essay Sur Le Commerce de Marseille**, 1842

Table 3.1b. Sulfur – Marseille. Imports in 1840.

Origin	Weight, tonnes
Two-Sicilies	16,649
Tuscany	277
Sardinia	61
Papal States	140
Austria	115
Total	**17,250**

Source: Jules Juliana, **Essay Sur Le Commerce de Marseille**, 1842

Between 1820 and 1824, British sulfur demand had averaged only 7,000 tonnes per year, but this quickly increased, after the salt tax was repealed. In the period 1825 and 1833, French demand was relatively steady, averaging 12,000 tonnes per year, but Britain roared past them as the major customer for Sicilian sulfur.

Sulfur was one of the few success stories in a seriously depressed area. Sicily was an economic basket case through most of the 19th century. The working-class lived in squalid conditions, employees of the sulfur industry among the worst.

Adding to the misery, major cholera outbreaks occurred on the island: in 1837, 1854 and 1855.

Le roi du sol in Sicily was under the control of the kings of Naples (also known as the Kingdom of the Two Sicilies): Ferdinand I (reigned 1759 to 1825); Francis I (reigned 1825 to 1830); Ferdinand II (reigned 1830 to 1859); and Francis II (reigned 1859 to 1860), the last king of the region before a united Italy. The state government looked down with contempt upon the Sicilians. They were a poor source of revenue and a burden to the state. The native Sicilians, in turn, resented the aristocratic rule of the "foreigners," the Neapolitan Bourbons.

The unification of Italy began when Guiseppe Garibaldi (1807 - 1882) landed with his one thousand "Red Shirts" in Sicily, on May 11, 1860. A sulfur tale is connected with the landing at Marsala. A telegraph message from Neapolitan defenders had been sent to the nearby military command at Trapani that two vessels had landed armed troops. Garibaldi's men took over the telegraph station moments later and sent their own message, which read that the previous message had been made in error. The two vessels were only carrying sulfur from Girgenti. The sulfur ruse had done the trick. The response from the army command was that the telegraph operator was a fool! The outnumbered freedom fighters went on to easily defeat the despondent Neapolitans. On September 7th, Garibaldi entered Naples, and Victor Emmanuel (1820 - 1878) was declared king of Italy two months later.

The new Italian state made some feeble attempts to improve the situation in Sicily without success (or much effort). The deplorable economic conditions continued to foster discontent and lawlessness. Shortly before Italy was unified, the mafia emerged in Sicily as a powerful and deadly organization. Kidnapping, robbery and murders increased. Protection rackets and extortion plagued Sicily. Feuds between rival "families" resulted in mass murders. To make matters worse, further cholera outbreaks took place in Sicily, in 1866, 1884/1885 and 1911.

3.3.2 THE ZOLFARE, THE SOLFATARI & THE CARUSI

The sulfur mines in Sicily were known as the zolfare. A major landmark in the history of the industry took place in 1750, when the mine at Floristella-Grottacalda opened in the Enna province of Sicily. This mine was owned by Prince Sant' Elia and the *gabelloti* (estate manager) was J. Trewhella and Company in the later 19th century. Floristella-Grottacalda was one of the best operated mines in Sicily. The mine was one of the first to adopt the new Gill furnace technology. The site produced sulfur for over two centuries and was not closed until the very end of the industry in 1984.

By the 1820's, there were already more than 200 small mines operating; among the largest were Floristella-Grottacalda, Villarosa, Santa Catalda, and Terra di Falco. Small mines continued to pop up across the central part of the island. By 1865, there were 615 sulfur mines, of which 378 were operating. A quarter century later, the number of mines had increased to 818, of which 581 were operating. Non-operating mines were not always depleted of sulfur; flooding was a common problem, which the operators simply walked away from instead of attempting to control the water. As time passed, mining became more expensive and dangerous as the miners had to go deeper and deeper. In 1870, the average depth of the Sicilian mines was 160 feet. By 1931, the average depth of the mines had increased to 625 feet.

Fires within the mines were a constant threat. Sulfur dust is combustible and any spark can cause a flash fire. For example, at the Trabonella mine, owned by Baron Morillo, there were serious fires in 1867 and 1911; in both incidents, about forty miners were killed. At the Cozzo Disi mine, opened in 1870, 89 miners were killed in a fire in 1916.

The miners were called *solfatari*. In 1890, the average Sicilian wage at the mines was equivalent to $0.60 (U.S.) per day, when the average wage in the U.S. was $5.00 per day. By then direct employment was 32,269 and another 18,000 indirect jobs were dependent upon the mines. General working conditions were deplorable. The emissions from the sulfur purification plants killed any local vegetation. The *solfatari* then could not even grow their own vegetables. Local supplies were usually purchased from the local *bettolino* (canteen), which was owned by the mine.

At the *zolfare*, child labor was used to help mine the sulfur. Boys, called the *carusi*, some as young as six years old, were assistants, more properly slaves, to the miners. The life of David Copperfield at Salem House was opulent by comparison. The *carusi* carried the ore out of the mine. They were usually sold to the miners, for a fee of several dollars (between 100 and 150 francs). If they could not afford to buy their freedom, they labored in the mines until they died. An article appeared in the *Times* (London) on November 27, 1893, condemning the treatment of the *carusi*. Even though illegal, children as young as eight years old still worked in the mines. They were paid 50 centimes per day, and were forced to sleep in the mines.

Among the early activists against the abuse of child labor and the inhumane conditions in the mines were: the priest Luigi Sturzo (1871 - 1959); the writer Giovanni Verga (1840 - 1922), who wrote a short story, **Rosso Malpelo**, outlining the horrors of the children working in the sulfur mines; the writer, Dr. Luigi Pirandello (1867 - 1936), whose father had owned a Sicilian sulfur mine and he wrote a novel about the carusi at Floristella-Grottacalda; the artist Onofrio Tomaselli (1866 - 1956); the composer, Alfredo Casella (1883 - 1947); and the famous American educator Booker T. Washington (1856 - 1915). Public outcry slowly forced a disinterested government to act. An early, but weak, step to improve the working conditions was made by the Italian government after the far-reaching report (**Sicily in 1876: Political and Administrative Conditions**) on the social, economic and political conditions of Sicily by Leopoldo Franchetti (1847 - 1917) and Sydney Sonnino (1849 - 1922); both were social scientists and members of parliament. Reluctantly, the minimum age for mine workers was increased by government decree:

1876	ten years old
1905	fourteen years old
1934	sixteen years old.

Women, too, worked in the mines. Another law, passed in the 1890's, stopped mothers from bringing their children into the mines.

3.3.3 THE DOPPIONI, CALCARELLA, CALCARONI & THE GILL FURNACE

There were two major operations within the sulfur industry: the mine itself, and the sulfur purification plant. The Sicilian deposits contained 12% to 50% sulfur, with the main contaminants being limestone and gypsum. Pure sulfur had to be extracted before sending the product to market. The simplest approach was to melt the sulfur. Fuel, though, was a luxury. Since sulfur burnt, it was used to melt itself! The earliest sulfur was recovered by placing the crude sulfur in conical mounds in an open-pit and covering it with earth. The bottom of the sulfur was set on fire, melting what remained on the pile.

During the Middle Ages, a more sophisticated, technology was the use of clay pots to distill the sulfur, as had been described by Biringuccio and Agricola. They were called *Doppioni*. There were, at least, two pots (*Doppioni* means "double"), connected by a clay pipe. The sulfur was heated in one and condensed in the other. The clay system was later replaced by iron retorts a meter high and a half meter in diameter. This technology was popular in the sulfur mines of the Romagna region of Italy, but never used in Sicily. The sulfur rights in Sicily were different from those on the mainland. In the region of Romagna, the land owners did not own the resources within the ground, which fell to the state. This regulation led to major companies controlling the sulfur in Romagna. By 1870, the major sulfur companies from this area were Società anonima delle miniere di zolfo di Romagna (formed in 1855), the Natale Dell Amore & C. and the Cesena Sulfur Company Ltd.

In the early 19th century, the process was "modernized" by placing the crude sulfur in open topped, sloped-bottom kilns called *Calcarella*. These were soon replaced by the larger *Calcaroni*. The major difference between the two was the addition of soil on top of the sulfur heap in the latter case, to control the combustion. The larger kilns, with diameters of thirty-five feet and a capacity of 28,000 cubic feet, would yield 200 tonnes of sulfur after burning for two months. The melted sulfur was collected in the sloped bottom of the *Calcaroni*. After several days, a plug was removed and the liquid sulfur was poured into moulds. The *Calcaroni* was less sophisticated and efficient than the Doppioni, but the former allowed processing on a much larger scale without the expenditure of much capital.

An improved chambered kiln was the Gill furnace, invented by Robert Gill (Italian Pat. No. 741, 1880), with two to six chambers. The first Gill furnaces were installed at the mines at Floristella-Grottacalda, Gibellini and Regalmuto. Many of the refineries were found at Catania, where there were seven facilities by the early 1890's. The largest produced 44 tonnes per day. In 1891, 75% of the sulfur in Sicily was produced from the *Calcaroni* and 17% by the Gill furnace. By 1903, the sulfur from the Gill furnace had risen to 55% and from the *Calcaroni* had dropped to 32%. Coke was often used as the fuel. Sulfur recovery improved to 65% to 75%. The *Calcaroni* and Gill furnaces remained the standard equipment in Sicily until the very end of the industry in the latter half of the 20th century.

The quality of the sulfur produced by the process was variable. The major grades of Sicilian sulfur were:
- *Gialla Superiore* (Best Yellow)
- *Gialla Inferior* (Inferior Yellow)
- *Buona* (Good) quality
- *Corrente* (Current) quality.

The purification processes were simple, but not without serious problems. About two-thirds of the original sulfur was lost by the crude open-pit method; the resulting sulfur dioxide escaped into the air. With the introduction of the *Calcaroni*, about one-third to one-half of the sulfur was still lost. While the burning of sulfur was more economic than using alternate fuels such as coal, the massive release of sulfur dioxide was not just wasteful but harmful to the workers and devastating to the local countryside. Thousands of tonnes of sulfur dioxide were released each year by the early 19th century. A story from the annals of the Royal Navy puts the situation in perspective. When the British held a protectorate over Sicily from 1806 to 1815, the navy brass had been so impressed by the destructive nature of the sulfur dioxide fumes from the Sicilian sulfur operations on the local flora and fauna that they began to investigate the military potential of the gas (a formal proposal was made to utilize the new chemical weapon during the Crimean War, but this was rejected). So much sulfur dioxide was released that large portions of the island were being sterilized.

Not until 1885 did the government pass regulations to restrict the processing of sulfur ores to only October to June, so that local crops could be grown. By this time, sulfur dioxide emissions had risen to hundreds of thousands of tonnes per year! The controls on refining forced another cog into the labyrinth of the already chaotic Sicilian sulfur system. For shipments in the off-season, massive inventories had to be kept in warehouses, where the product was held on consignment. The warehouses became the link between the *gabelloti* and the merchants. Speculation on the value of the inventory got out of control. Some warehouse operators started to manipulate the market; others went bankrupt.

3.3.4 THE GABELLOTI

Most of the economic benefits of the sulfur industry were drained out of the island. Taxes on sulfur went to Naples. The operating companies at the mines, called the *gabelloti*, had to pay fixed lease agreements to the numerous absentee landlords, usually from mainland Italy. The aristocratic Italian owners took sulfur for payment in kind, adding further twists to the complicated supply chain. The sulfur mine rent was about 20% of production. The contracts with the *gabelloti* also required a minimum production, regardless of demand.

The infrastructure system in Sicily was disastrous. The basic problem was the separation between ownership, operations and marketing. There was no system but a bunch of independent factions, each taking care of their own selfish interests. The end result was that there was no incentive for anyone to make investments into the business. Thus, the industry remained primitive and backward. Worsening the situation, the economy of Sicily became more and more dependent on the sorrowful mines, both for employment and income. On the rare occasions when attempts to modernize the industry and improve efficiency (i.e., job losses) were proposed, they were met with violent demonstrations and strong resistance from local residents and political leaders. This Sicilian sulfur pit trapped the industry, and the island economy, in a state of mediocrity. When serious competition later evolved and their monopoly was broken, only government intervention kept the obsolete industry afloat.

3.3.5 THE BRITISH WINE MERCHANTS

The Sicilian industry was composed of numerous independent, inefficient operations, whose commercial success often depended upon marketing consortiums to control the international price of their product. Any real profits from the sulfur industry of Sicily were reaped by foreign bourgeoisie, especially British, who controlled the exports from the island.

The British dominance of the Sicilian sulfur trade (and other exports) can be blamed on Napoleon. In 1799, Napoleon had conquered Naples, but Ferdinand continued to rule from Sicily, being taken there for refuge in the flagship of Admiral Nelson. His rescue and subsequent protection by the British led to strong commercial ties between the two countries. Between 1806 and 1815, the *decennio inglese* ("English decade"), the British held a protectorate over Sicily. A French invasion was prevented, but replaced with an invasion of British merchants; up to twenty of them had come to Palermo and another forty to Marsala. In 1815, Ferdinand was returned to Naples by the British, and the following year, he officially became king of the Two Sicilies. The British government received a most-favored-nation agreement from Ferdinand I, in the Anglo-Neapolitan treaty of September 26, 1816. This treaty virtually converted Sicily into a British trading colony.

The lucrative wine trade had first brought them here, but they soon diversified into other goods, especially sulfur. The pioneering British wine merchant was John Woodhouse Sr., who arrived long before the trouble with France, in 1770. Three years later, he opened a winery in Marsala. His son John Woodhouse Jr. (1768 - 1826) arrived in 1787 and between them they created a global market for fortified-Sicilian wines. Their success was soon surpassed by Benjamin Ingham (1784 - 1861). He had been among the first to take advantage of the British protectorate, arriving in Sicily in 1806. At first, he was representing his family's firm Ingham Brothers & Company, but in 1809 formed his own enterprise. Three years later, he went head-to-head against Woodhouse in the wine business. After several years of struggle, Ingham's trade surpassed that of Woodhouse. B. Ingham and Company became Ingham, Stephens and Company (with Richard Stephens), and then Ingham, Whitaker and Company [with Joseph Whitaker (1802 - 1884)] in 1868. The firm became involved in the sulfur business, using it as ballast for their merchant ships. Near 1830, with Prince Emanuele Pantelleria (? - 1848), Ingham was operating sulfur mines at Girgenti.

There was no direct relationship between the sulfur and wine businesses. The merchants were essentially logistics experts with international marketing connections. They knew best how to transport goods out of Sicily to England, the U.S., or other parts of the world more efficiently than anyone else. Having access to the greatest merchant fleet in the world didn't hurt either. Since its modern inception, logistics was a pivotal aspect of the sulfur business, a factor that would only increase in importance over later decades.

One of the first sulfur merchants in Sicily was not British, but American, Benjamin Gardner (? - 1837) of Boston. He arrived in Sicily in 1819. Gardener may have first been attracted to Sicily by a diplomatic mission. Just after he arrived, Henry Preble (1770 - 1825), the American consul in Sicily resigned in 1820, and Gardener was made acting consul. On December 23, 1825, Gardner officially replaced Preble. His appointment, approved by John Quincy Adams, had been nominated by the famous Daniel Webster (1782 - 1852). Webster, a prominent lawyer at the time in Boston, must have known Gardner before he moved to Sicily. During the early boom years in sulfur, Gardner became partners with James Rose (? - 1868) of Britain, forming Gardner, Rose & Company. Their mines were found at Lercara, forty miles from Palermo. During a visit to the company mines in 1876, the youngest son of James Rose, John Forester Rose, was kidnapped by the legendary Sicilian bandit, and early leader of the Sicilian "mob," Antonino Leone; Rose was later released unharmed upon payment of a ransom. The firm failed shortly afterwards, after a collapse of sulfur pricing.

George Wood was another Marsala wine merchant that had entered the sulfur business. The Wood family had come to Sicily in the early 1800's. In 1882, Wood & Company was purchased by Ingham-Whitaker. A sulfur trader not involved in the wine industry was Morrison & Company. The Morrison family had also been early merchants in Sicily, dealing mainly in licorice paste. The business had been founded by James Morrison (1789 - 1873). Another major sulfur trading house was owned by Robert Frank of Licata, near Girgenti. By the early 1830's, about twenty British trading firms had set up in Sicily, and British entrepreneurs exploited the resources of the island.

While the sulfur trade was dominated by British interests, one Italian businessman stood out, the legendary Vincenzo Florio (1799 - 1868). He moved with his father, Paulo, from Bagnara Calabra in Calabria to Sicily, where Paulo opened a pharmacy. Vinzenzo, too, entered the wine business, in 1832. With his son, Ignazio Florio (1838 - 1891), they formed I. & V. Florio. Eventually, this Sicilian-based business would surpass those of his friend Ingham, Gardner and the others. Florio and Ingham became business partners. In 1841, they purchased a sulfuric acid business in Palermo, Chimica Arenella (later becoming the citric acid plant of the German firm Fabbrica Chimica Italiana Goldenberg in 1912; plant closed in 1965), which had been founded by Augustin Pourri. The previous year, they had formed a shipping business together. Another successful Italian trading firm was founded by Sebastiano Lipari, who was the consul from Piedmont. During the middle of the century, the leading merchants were Ingham-Stephens, Woodhouse, Florio, Lipari, Gill & Corlett, and Wood & Company.

[The grand-old wine merchants of Sicily began to fail by the end of the century. Their markets had declined and competition had increased. The latest generation did not have the entrepreneurial spirit of their grandfathers, and their free-spending ways drained their inheritances. During the 1920's, the government was not particularly friendly towards them. Benito Mussolini (1883 - 1945) had become Prime Minister on October 30, 1922. His family had been involved in the sulfur business; his uncle Alcide Mussolini (~1865 - 1929) had been a foreman in a sulfur mine near Forli. In 1928 and 1929, the firms of Ingham-Whitaker, Florio and Woodhouse ended up in the hands of their competitor Cinzano. The three were combined to form a new firm named Societa Anonima Vinicola Italiana.]

3.3.6 THE SULFUR WAR OF 1840

The 1830's were the most turbulent decade in the history of the sulfur industry, and the British merchants were smack in the middle of it. Sicilian sales grew at rates never seen before in any industry, largely due to the escalating number of Leblanc plants in England. Sicilian sulfur production had reached 31,700 tonnes by 1830, an astronomic demand for the times, with most (90%) product being exported to Britain and France. Market forces, though, were just beginning to strain the sulfur resources of Sicily. Throughout the 1830's, French demand continued to grow at modest rates (see Table 3.1), but the gap with Britain widened. Britain, alone, was soon importing more than 32,000 tonnes per year! Operators and merchants in Sicily had done very well ramping up production and logistics to keep up with the unprecedented rapid growth in demand.

Such a rapid rise in demand is often accompanied by a parallel rise in pricing, which is resisted by the market place. The degree of resistance is directly proportional to the magnitude of the increase. What started off as a market battle almost turned into a real one! An ironic side note was that both sides in the initial trade war were British, one controlling the sulfur trade and the other, the largest consumers in the world, the Leblanc producers of England. What really pushed the British towards war was when French interests had taken over the Sicilian sulfur trade from their countrymen.

The British merchants who controlled the Sicilian sulfur trade got greedy. Pricing had been steady at the equivalent of $20 (U.S.) per tonne level. In 1831, the price was raised to $70 per tonne! Since they controlled the global trade, what choice did the consumer have; pay up or close down. British or not, this was business. The increase had taken place just after Ingham had entered into the sulfur mining business, which was probably no coincidence. Muspratt had just consolidated his operations at Newton. He and the other Leblanc industrial leaders in England were not novices in dealing with ruthless business practices, many of which they had invented themselves. Their countrymen in Sicily had grabbed the Leblanc bull by its proverbial sulfur horns. The strategy of the Leblanc industry was to purchase in excess of demand to build stockpiles. Sicilian exports shot up to 72,000 tonnes the following year! New record pricing was reached at $88 per tonne. The merchants were laughing all the way to the bank...for now. The sulfur-consuming industries were taking a terrible hit on costs, but their strategy was falling into place. While 1832 was an incredibly exciting year for the British sulfur merchants, the consumers turn would shortly come. In 1833, purchases tumbled as the major customers stopped buying, letting their inventories draw down. Sicilian exports plummeted and pricing collapsed to $15 per tonne. The Leblanc producers wanted to protect themselves in the future. Tennant, Muspratt and other Leblanc companies invested in their own Sicilian sulfur mines in 1835.

However, the sulfur schism of the early '30's was only a skirmish of a more serious trade dispute that was about to begin. As the yellow dust was just settling, an enterprising team of entrepreneurs from Marseilles saw a business opportunity among the turmoil. In 1836, Aime Taix, and his partner Arsene Aycard, sought an exclusive license for the sulfur trade from Ferdinand II; in other words, the two French businessmen wanted to take the business away from the British. The proposal was brash and daring, since the business had belonged almost exclusively to the British for the past two decades. The Sicilian industry, though, was in deep trouble, worse than normal, as pricing had remained depressed. The offer by Taix may have opened the eyes of the Neapolitan government to the true value of their Sicilian assets. The proposal also affirmed that Naples was getting the short end of the sulfur stick in the deal with the British. The mega-project of Taix was no simple distributor agreement. Major investment would be made in Sicily and millions of dollars in revenue would be sent into the court coffers. Such multi-million dollar deals were rare. Ferdinand was interested, but was not yet ready to risk the dangerous transition. The British sulfur traders were fuming over rumors of a deal, but the government reassured the British ambassador that the offer was being rejected...for now.

The daring partner was Taix. Although he was paddling into dangerous international political waters, there was a crack in the sulfur (an inherent problem of the element) business that he was going to exploit. The royal court in Naples had, at least, listened to him, and his plans had sparked some lively discussions. Taix was not deterred and came back again in September of the following year. The British were again not amused. After a terse letter from Lord Palmerson to Ferdinand, Taix was rebuffed once more. Meanwhile, the sulfur business continued to surge ahead. In 1837, British demand was back to a respectable 37,486 tonnes, with France taking half this amount. Global demand for Sicilian sulfur was more than 60,000 tonnes. By the end of the year, the British merchants in Sicily may have felt that the Taix nuisance was behind them, but they were wrong. Taix was not going to let this opportunity slip away.

The persistent Taix came back yet again to Naples and was third time lucky. His constant badgering and aggressive and persuasive arguments had at last won over Ferdinand, British protests or no. The French promoter had essentially bribed the government with all sorts of promises of investment and revenue. Taix, though, did not have much to back up his promises, having little capital or investment funds. Ferdinand may have felt that he had much to gain and not much to lose, so he sanctioned the bold venture of M. Taix.

Ferdinand abruptly signed a verbose royal decree on June 27, 1838, which was released on July 4th, forming the new joint venture company, Taix, Aycard & Company (TAC; the British press dubbed the new firm, the Sulphur Company.) The decree began (von Raumer, F., **Italy and the Italians**, Vol. II, p. 312, Henry Colburn, London, 1840):

For the benefit of our beloved subjects, in order to pay debts in Sicily, to alleviate burdens, to diffuse great wealth, and to call forth public works, which the island has such need of, a contract is concluded (without listening to plans of rights and privileges), Taix, Aycard and Co. for ten years, to the following purport:

1.As the great production of sulfur is the cause of every calamity in Sicily, the same shall be reduced from 900,000 quintals to 600,000 per annum, consequently diminished one-third.
2. The average produce from 1834 to 1837 shall determine the quantity of the two-thirds, beyond which no sulfur shall henceforth be allowed to be raised. *[This clause was later replaced by an allocation set by the government as the producers inflated their previous sales to increase their quotas.]*
3. The price at which the company buys and sells shall be officially fixed.
4. It pays to the king 400,000 Neapolitan ducats per annum.
5. The proprietors have full and unlimited liberty to sell their sulfur to whomsoever they please, and to send it whither they will, in case they do not choose to depose of it to the company.

Other terms of the ten-year contract were that TAC would:
- purchase the sulfur at 21 to 25 carlins per cantar, according to quality (before the TAC deal, the purchase price was half this amount).
- sell the sulfur at 41 to 45 carlins per cantar.
- pay 4 carlins per cantar to compensate the mines for limiting production to two-thirds of their annual production (based on the average production of 1834 to 1837).
- pay 66 2/3 grains per cantar to the government, in addition to the annual fee of 400,000 ducats.
- pay one-third of the profits to go to the government if TAC sold more than 600,000 cantars of sulfur.
- maintain a minimum sulfur inventory of 150,000 cantars.
- spend £10,000 per year to build new roads in Sicily.
- export one-third of the sulfur in Neapolitan vessels, which had been dominated by British merchant ships.
- build a sulfur refining plant at Girgenti, and, within four years, a Leblanc plant.
- invest 800,000 ducats directly. Another 400,000 ducats were to be raised through a share offering to the mine owners and citizens of the country. The Neapolitan government, itself, was going to invest 600,000 ducats

into the venture. The total investment in the new company was to be 1.8 million ducats (£288,000 or 7.2MM FF in 1840). Funds for the project were to be provided by J. Lafitte and Company.

The mixture of obsolete units above is confusing. The definition of the cantar (or quintal, Arabic form of the word) is especially difficult to convert into modern units, since it varied by country (in Naples = 196 lbs.). The basic currency of Naples was the grain, with the following conversions:

$$1 \text{ carlin} = 10 \text{ grains}$$
$$1 \text{ taris} = 20 \text{ grains}$$
$$1 \text{ ducat} = 100 \text{ grains}$$

In 1850, one ducat was equal to $0.80 (U.S.). Using this exchange rate and the cantar at 196 pounds, the above data can be translated into U.S. dollars of the day and tonnes:

1. Total capital investment = $1.44 million
2. Fixed revenue fee to Naples = $320,000 per year
3. Variable royalty fee to Naples = $6 per tonne (~$320,000 per year)
4. Selling price for sulfur = $40.50 per tonne
5. Sulfur production would be reduced from 80,000 tonnes to 53,350 tonnes per year
6. Compensation to mines for reduced production = $3.60 per tonne (~$96,000 per year)
7. Minimum sulfur inventory = 13,336 tonnes.

Ferdinand was willing to offend the greatest sea power that the world had ever seen over sulfur. With this decree, the government had frozen out the British merchants in Sicily and placed sulfur under allocation and price-controls. Since *the great production of sulfur is the cause of every calamity in Sicily* was it not the duty of government to protect its people? The official explanation for the agreement was to prevent depletion of the sulfur reserves and to provide reasonable return to the poor Sicilians. In a further weak attempt to dampen the impact of the document, the term monopoly was deliberately avoided. Not everyone had to sell through TAC, but, if they did not, they had to pay TAC 2 ducats per cantar ($18 per tonne). Such penalties prevented any real competition to TAC. Thus TAC was a monopoly in all aspects but name only.

The new deal was implemented with amazing speed. The British merchants were caught off guard, as the new sulfur deal took effect in only a few weeks, on August 1, 1838. The decision came so quickly that twenty-four British merchantmen that had been chartered to pick up sulfur had to be turned away at the port, because they arrived after the August 1st deadline.

While against the spirit of the treaty of 1816, Ferdinand was arguably not in official violation of a two-decade old agreement with the British; he was simply taking control, through TAC, of resources that belonged to his government. He was stretching the relationship. In any case, how much longer did his country have to pay back the British for their help against Napoleon? The TAC agreement, if carried out to the letter, would have greatly benefited Sicily and Naples, compared to the British deal. At this stage, Ferdinand had been wise in his sulfur strategy. He expected the British to rant and rave. What could they really do? Would the British fleet be sent against his country over a silly dispute over sulfur?

The new price and allocation were a rude shock for the market. The alkali industry (i.e., Leblanc industry), the largest customer of Sicilian sulfur, was again threatened by the high pricing and, more importantly, a cut back in supply. Within a year into the TAC deal, the situation began to deteriorate, especially for the British, and even for the French and the Sicilians themselves. In 1839, trade with Britain fell drastically. Between 1838 and 1839, total sulfur imports from all sources into Britain dropped from 44,000 tonnes to 22,000 tonnes; of this latter amount, only 5,000 tonnes had come from Sicily. The supply restrictions had forced British consumers to even import one shipload of sulfur from Iceland. The year 1838 had been an exceptional time for French imports, reaching over 36,500 tonnes. French trade also dropped, but not as drastically, from a value of 3.7 million FF in 1838 to 2.5 million FF, in 1839. In 1840, they dropped further to 2.1 million FF. With sales falling, overall Sicilian sulfur sales were also depressed, despite the hefty price increases: 9.4 million FF in 1838, falling to 3.4 million FF in 1839. During the period August 1838 to February 1840, TAC exported 61,245 tonnes of sulfur from Sicily (see Table 3.2).

Table 3.2. Sulfur – Sicily. Exports in tonnes

Importer	8-1-1838 to 2-21-1840	2-22-1840 to 12-9-1840
England	31,850	4,919
France	27,087	9,934
Netherlands/Belgium	1,295	508
Russia	932	0
Germany	82	0
Malta	0	90
Miscellaneous	0	32
U.S.	0	293
Total	**61,245**	**15,775**

Source: Jules Juliana, **Essay Sur Le Commerce de Marseille**, 1842

British politicians jumped on the anti-Naples band-wagon as their economy and darling Leblanc industry were being abused by this savage foreign government. John Singleton Copley (1772 - 1863), the Baron of Lyndhurst, stood up in parliament and claimed that the French monopoly in Sicily was costing the British £1,000 per day! A number that was subsequently widely quoted. The monopoly to

TAC allegedly threatened the entire economy of Britain. The government of William Lambe, Viscount Melbourne (1779 - 1848), had to do something.

During the first year of TAC, the British government was relatively passive. The British minister in Naples, Kennedy, complained to the royal court in July 1839, a year into the deal. Ferdinand was unnerved by the terse letter. A meeting was held between Taix and Ferdinand on August 24th. Official documents state that Ferdinand was going to cancel the contract but that Taix, as determined as ever, argued convincingly against it. In September 1839, L. M'Gregor, the Secretary of the Board of Trade, was sent as Commissioner to Sicily to evaluate the situation and hold discussions with the Neapolitan court. He returned, and on October 25th, he made the startling recommendation that duties be placed against Sicilian sulfur to ban their importation. Obviously, M'Gregor was oblivious to the fact that no other country could meet British demands. The recommendations were not carried out. M'Gregor had pressed the Neapolitan court and was told that the monopoly would be removed by the end of the year. The New Year came and went, and TAC still controlled the Sicilian sulfur. In February 1840, Palmerston (1784 - 1865) sent another threatening note to Ferdinand. Prince Cassaro, the respected minister of foreign affairs, responded on February 23rd that the king had agreed to again cancel the deal with TAC. A strange aspect of the note was that Cassaro asked Kennedy to keep the information confidential so that "we should not get into difficulties with M. Taix." Ferdinand, though, changed his mind shortly afterwards and told the British minister, William ("Sleepy William") Temple (1788 - 1856), who was the younger brother of Lord Palmerston (Henry John Temple), that the TAC contract would remain in force. Prince Cassaro objected and was dismissed by Ferdinand on March 21, 1840. William Temple's role throughout this dispute is puzzling. While he was the British ambassador to Naples, he remained in Britain from mid-1838 until the end of February 1840.

Since court diplomacy had not worked, the British notched up the pressure to gunboat diplomacy. Admiral Robert Stopford (1768 - 1847) was ordered to make the Mediterranean fleet ready for war in March and was officially ordered into action on April 1st. The Sulfur War was becoming a pan-European affair:
- Sicily, the source of the sulfur;
- Naples, who changed the business structure of the industry;
- France, the home of the company who now controlled the industry;
- Britain, the largest trading partner and the one who used to have the monopoly on sulfur trade.

All this fuss over sulfur! Naples sent 12,000 troops and an artillery division to Sicily and prepared for war. If the situation continued to escalate, there was the chance of a wider conflict drawing in the super-powers of Europe. The aggressive stance of the British was causing waves across the Continent. Rumors spread that

Britain was taking advantage of the weak country, and that their real intentions were to occupy Sicily to enhance their control of the Mediterranean. Even Palmerston felt that things were getting a little out of hand. Fortunately, this was a war in name only. Not a shot was fired. Not a person was harmed, besides a few hurt egos.

Calmer heads prevailed. On April 10, 1840, Palmerston met with the French ambassador Francois Guizot (1787 - 1884). The former seemed uneasy that such a relatively minor issue had escalated to military action. Palmerston asked Guizot to mediate a settlement. Two days later, the French government of Louis Philip (reigned 1830 - 1848), the son of the patron of Leblanc and now the king of France, agreed to the request. On April 17th, the British fleet had arrived at Naples. Besides a few ships sailing the Sicilian flag being seized by the *H.M.S. Talbot*, the Sulfur War was over before it began. Ferdinand accepted the mediation by the French on April 26th, and the following day, the British fleet stayed hostilities. On July 21st, the monopoly of TAC had been revoked by Ferdinand and was returned to the British, and on November 17, 1840, the trade war that was becoming a real war officially ended.

Although their grandiose sulfur plans were now dashed, the business culprits behind this conflict were rewarded for their troubles. Compensation to Taix and Aycard for the breached agreement was paid by the French government. British merchants who were hurt by the sulfur embargo were reimbursed by the Sicilian government. The compensation was determined by an international tribunal from Britain, France and Naples. The two British commissioners were the seasoned diplomats Woodbine Parish (1796 - 1882), who became Chief Commissioner to Naples from 1840 to 1845, and Stephen Henri Sulivan, the nephew of Palmerston. The two commissioners from Naples were M.A. La Rosa and G. Bongiardino. The chairman was the career diplomat Alexandre Louis Thomas Lurde, Comte de Lurde, of France. On December 24, 1841, the final compensation list was issued (see Table 3.3). The total compensation was only 121,454 ducats or £21,300. The Commission had rejected most of the claims, which had totaled 373,978 ducats or £65,610. Considering the small amount of damage by the ruthless monopoly of TAC, one wonders what all the fuss was about. According to the earlier war-mongering estimates of the Baron of Lyndhurst, the damage should have been over £500,000. The total volume of sulfur claimed was only 366,420 quintals, or 32,577 tonnes, the bulk of which came from Wood & Co. The companies receiving payment, included sulfur producers, traders and consumers. The only sulfur producer/traders in the list were G. Wood & Co. and Morrison & Co. Notable by their absence were Ingham, Gardner & Rose, Muspratt and Tennant, who did not file claims.

Table 3.3. Sulfur war compensation

Company	Accepted claim, tonnes	Compensation, £
Mine owner		
G. Wood & Co.	16,538	12,512
Morrison & Co.	2,492	2,804
Trader		
Mathey Oates & Co.	56	52
Samuel Lowell	disallowed	disallowed
Consumer		
Prior Turner & Co.	1,274	3,508
W. Leaf & Co.	2,073	1,578
Frank Ball	198	263
T.& R. Sanderson	175	521
Thurburn Rose	52	61
Total	**22,859**	**21,307**

[A doctoral dissertation on the Sulfur War was prepared by Dennis Thomson of Michigan State University. The Sulfur War is briefly mentioned in a satirical poem on Lord Stanley by Thomas Moore called **Thoughts of Mischief**.]

The sulfur wheeling and dealing were not quite over. As the monopoly was canceled, the government of Naples established a duty of 20 carlins per cantar (£4 10s per ton in 1841). The Neapolitan king had, unwittingly, fostered the cause of pyrites. On October 21, 1841, John Goodwin, the Consul-General, pleaded with Temple to get the duty removed; a petition was attached from Sicilian mine owners and merchants. The only English name was that of Ingham. The others had refused to sign because it was a "Sicilian" petition.

3.3.7 THE POST-SULFUR WAR ERA & THE RISE OF PYRITES

This period inaugurates a paradigm shift in the sulfur industry, with the rise of pyrites as a competitive product to elemental sulfur for sulfuric acid production. Incredibly, pyrites became a more important sulfur source than elemental sulfur itself! This absurd change had been caused by the instability of supply during the Sulfur War.

The pyrites industry is a marketing anomaly that by right should have only developed as a niche supplier, at most. Pyrites are actually a poor source of sulfur, having a low sulfur content (<50%), resulting in high transportation costs, and producing a problematic solid waste product. Despite these competitive disadvantages, for a period of one hundred years (1841 to 1940), they held significant global market share. Overall, Rio Tinto and the other pyrites companies must have been exceptional suppliers, especially considering their product drawbacks and the long supply lines, or conversely, the sulfur suppliers were...

The sulfur-to-pyrites switch is one of the earliest examples of an industrial shift in technology driven by supplier ineptitude. The incident is also the first to demonstrate the potent power of industrial R&D when mobilized in a common cause; fifteen patents were issued on the production of sulfuric acid from alternate sources, especially pyrites. The incident became an early "case study" on harnessing science and technology to overcome threats to industry: such commentaries were presented by Justus von Liebig in 1843 (**Letters of Chemistry**) and Pierre-Joseph Proudhon (1809 - 1865) in 1847 (**The Philosophy of Misery**).

Market pull and supplier push are well known market forces, but the process was all screwed up in this case. There was no problem with the market pull part. It was the supplier push part that was confused. The pyrites companies never truly took market share away from elemental sulfur; the latter pushed the market to them. The Sicilian industry had only itself to blame for driving consumers to this alternate source of supply. Sulfur as the preferred raw material for sulfuric acid could only be sustained as long as it maintained its economic advantage and there was security of supply. Failure of these principles was the catalyst that transformed pyrites into a global source of sulfur. Without such an event, pyrites may never have amounted to much in the sulfur world. Once in their iron fists, the customers, who always have long memories, were not easily converted back to elemental sulfur. The pyrites companies, of course, gladly took advantage of this strange situation (see above).

The decade, after the Sulfur War, had started off badly as a general depression had hit in the early 1840's. Overall, sulfur trade began to pick up in 1844, even though pyrites had already taken a large share of the market. The sulfur duties had not helped. In August 1845, Ingham went to Naples to discuss the duty with Ferdinand.

The industry was not down for long and was booming again during the 1850's (see Table 3.4). The growth was exceptional, especially in light of the major loss of market share to pyrites. The global sulfur trade surpassed 100,000 tonnes for the first time in 1854. German and Swiss trading companies had become established in Sicily. In 1857, the largest of these new foreign firms, Fratelli Jung started trading in sulfur. Another new sulfur trader was Kayser & Kressner.

Table 3.4. Sulfur – Sicily Exports 1851 to 1860

Year	Amount, tonnes
1851	86,150
1852	88,939
1853	97,268
1854	111,993
1855	101,393
1856	121,550
1857	125,987
1858	163,629
1859	152,487
1860	137,745

Source: Kingliest, C.T.: **The History, Products and Processes of the Alkali Trade: Including the Most Recent Developments**, and others.

With the unification of Italy, the government began to provide consistent regulation across the country. For the mining industry, in which sulfur was a major part, the body responsible for this task was the Consiglio e Corpo Reale delle Miniere, formed in 1865. Sulfur markets continued to grow. In 1868, 184,000 tonnes of sulfur were shipped from Sicily, valued at £1.5 million.

Sulfuric acid demand surged ahead at astronomic rates, but no new plants in Europe were using elemental sulfur. All new plants were on pyrites, and the established facilities switched in the following decades. By 1880, the elemental sulfur market in Europe for the manufacture of sulfuric acid had disappeared! The resilient demand for sulfur, though, was almost oblivious to these changes and continued to surge ahead. By the mid-1880's, Sicilian exports had double in two decades to 350,000 tonnes.

The fastest growing sulfuric acid market, the U.S., remained with elemental sulfur until the end of the century. While Sicily was losing most of its sulfuric acid market in Europe, U.S. demand was racing to replace it. In 1892, Sicilian exports to the U.S. alone were 89,000 tonnes. Later, the U.S. sulfuric acid industry also switched to pyrites. There had been no American duties on elemental sulfur, but there was on pyrites, which were removed in 1890. In 1895, 75% of U.S. sulfuric acid was still being manufactured from elemental sulfur, but had dropped to only 16% in 1901.

The luck of sulfur continued though. The loss of another major market did little to dampen the unstoppable growth of sulfur. By then, new major markets for elemental sulfur had been established, especially for the production of sulfite pulp (uses sulfur dioxide from the burning of sulfur), pesticides (for grapes) and rubber manufacturing (vulcanizing).

A new source of sulfur came on in Europe. For the first time, Sicily had some competition for their war-rousing element. A technology to recover sulfur from waste of the Leblanc process was developed. The new source had only a major impact on pyrites producers, since Britain was no longer using much elemental sulfur from Sicily directly. Carl Friedrich Claus (~1829 - ?), working in London in 1883, patented (Br. Patent 3608, 5070, 5958, 5959) the furnace to carry out the yellow transformation. The Claus discovery led to an immediate drop in the price of pyrites to the industry by 50%. With the lower pricing, it was not economical to recover the sulfur, so pyrites demand was, at first, largely unchanged. In 1887, Alexander M. Chance (~1845 - ?) and C.F. Chance (Br. Patent 8666) modified the process using only waste streams from the Leblanc process to produce the sulfur, still using Claus' technology. The source of the sulfur was the by-product calcium sulfide from the Leblanc process:

$$CaS \text{ (calcium sulfide)} + CO_2 + H_2O \rightarrow H_2S \text{ (hydrogen sulfide)}$$

$$H_2S + heat \rightarrow S + H_2O$$

Previously, the waste calcium sulfide had been heaped in giant mounds by the Leblanc plants. These stinky piles slowly released hydrogen sulfide into the air.

By the early 1890's, Britain had become the second largest sulfur producer in the world, by using the Claus-Chance process. In 1893, British sulfur production by this technology was 35,000 tonnes. The process would be famous again, when it became the standard technology for converting hydrogen sulfide in sour gas and oil into recovered sulfur. Then it would only be called the Claus process, since the Chance modification was applicable only to the Leblanc process.

3.4 THE CARTELS

While the new British sulfur posed some problems, a more serious crisis hit the Sicilian producers between 1893 and 1896. Pricing had increased in 1890 and 1891, which led to over production in 1892. Making matters worse for the industry, high taxes were taking what slim profits were being made. The industry was inflicted with varying regional taxes, plus a national export tax. Both mine operators and sulfur merchants began to fail. Riots broke out. The government, being pulled by several opposing forces, was paralyzed by the situation. The industry was in a crisis like it had never seen before.

TAC had been the first cartel that would control the marketing of Sicilian sulfur. After TAC, though, a new cartel was not formed for half a century. Afterwards, they became the standard way of doing business within the Sicilian sulfur industry.

3.4.1 ANGLO-SICILIAN SULPHUR COMPANY (ASSC): 1896 TO 1906

The British, as well, were not too pleased by the collapse of the sulfur industry, now that they were becoming major producers themselves. The United Alkali Company, itself a cartel of British Leblanc producers, formed an alliance with the leading Sicilian trading firm I. & V. Florio, now under the control of Ignazio Florio Junior (1868 - ?). The choice of Florio by United Alkali demonstrates the success of his company and the faltering position of the established British merchants in Sicily. Before ASSC could be formed, the government would have to make some concessions. On July 27, 1896, just days before the official launch of the ASSC, the government repealed the sulfur taxes, after being lobbied by Ignazio. The involvement of United Alkali in the operations of ASSC is not well documented. Considering the background of the British firm, they were likely not silent partners.

The new joint venture, called the Anglo-Sicilian Sulphur Company, was formed on August 1, 1896, with a five-year renewable term. They controlled through contracts two-thirds of Sicilian exports. While ASSC kept pricing firm, they were not in complete control of exports. Another one-third of the Sicilian industry, representing more than one hundred thousand tonnes of sulfur, was independent of the ASSC. The non-consortium members could take advantage of their discipline, but did not have to pay any of their commissions.

The U.S. market was especially important to the new firm, where they sold through some of the best known agents in the country, including Parson & Petit and C. Talamo Rossi, both of New York City, and Alec C. Fergusson of Philadelphia. Sulfur sales surged under the ASSC. By the end of the century, sales had grown to 424,000 tonnes, with a value of £1.7 million. ASSC increased pricing from 40 lira per tonne to 100 lira per tonne. An American traveler to Sicily in 1899 praised the increase. He had been previously in the country in 1895 when people were starving and there were bread riots. Strikes had shut down the mines. The stability of ASSC and the higher pricing had brought normality to the island. Ignazio Florio and the ASSC were sulfur heroes in Sicily.

In the midst of all this success, the Sicilian sulfur empire was on the brink of collapse. What the ASSC did not realize was that a golden horde was massing to overthrow Sicily as masters of the sulfur world. This bayou brimstone came from the marsh lands of Southern Louisiana! At the turn of the century, the new formidable foe disrupted the delicate balance established by ASSC: a new massive mine using the revolutionary Frasch process for mining sulfur had been opened by Union Sulfur. Caught between lower demand and the domestic producer, Sicilian exports to the U.S. collapsed: in 1904, they had been over 100,000 tonnes; two years later, they had almost disappeared, being only 7,000 tonnes! Inventories rose at an alarming rate in Sicily, from 306,000 tonnes in 1901 to 558,000 tonnes in 1905 (see Table 3.5). These inventories were equal to the annual world demand! Not only was the global dominance of Sicily over, in a couple more years, they were relegated to a secondary supplier. Such a swift collapse of an international chemical business on this scale had never been seen before and has seldom been seen since.

Table 3.5. Sulfur – Sicily from 1895 to 1906 in 1000 tonnes

Year	Production	Total sales	Sales to US	Inventories
1895	353	364	101	
1896			132	245
1897	475	425	123	
1898			143	
1899	537	424	134	
1900	535	560	166	
1901	554		150	306
1902	531		177	
1903	545		157	
1904	519	508	106	
1905	560	476		558
1906	492	322	7	

Source: Haynes, **The Stone That Burns**, and others

Not knowing any better, the mighty ASSC was going to show the upstart Americans who was running things. Not realizing that they had a yellow tiger by the tail, ASSC first warned Union Sulfur to stay out of Europe and also arrogantly demanded a share of the U.S. market. What the Sicilians did not yet appreciate was that the manufacturing-cost advantage of Union Sulfur was greater that the logistic-cost advantage of the Sicilians shipping to mainland Europe. No competitor could withstand the competitive edge of Union Sulfur, especially not the flimsy structure of the Sicilian sulfur industry.

Union Sulfur did not take kindly to the threat from ASSC and responded with bold action. In 1904, the Americans dropped the price of sulfur from $26 per tonne to $15 per tonne. Then, the dynamic Frasch, the president of Union Sulfur, met with the directors of ASSC in London. He frankly told them that he had to have the American market for himself. Considering the meeting was in London, members of United Alkali must have been there as well, although there is no direct documentation of this. The ASSC responded that the entire thing was "American humbug," and that it was an "American swindle." The colorful meeting broke off with no agreement. ASSC soon realized their folly when a 3,000-tonne load of American sulfur showed up in Marseilles in November, and then the *Mineral Industry 1904* reported that Frasch claimed that his costs were so low that he could deliver sulfur to Sicily cheaper than the Sicilians could make it. In 1905, they sent one of their directors, L. Dompe, to spy on the new competitor. He confirmed their worse fears. The ASSC then reached an agreement with Union Sulfur (Frasch-Sicily Agreement #1 [FSA-I]). They agreed to limit exports to the U.S. to 75,000 tonnes, if Union Sulfur stayed out of Europe. Under the circumstances, the terms were generous to the Sicilians. Later, it was claimed that this had been the case [C.F. Chandler, *J. Ind. Eng. Chem.*, **4** (2), p. 134 (1912)]:

> Fortunately the Company [Union Sulfur] is owned by a few broad-minded and large-hearted men, who could not be induced to bring starvation and ruin upon the two hundred and fifty thousand people dependent upon the mining of sulfur in Sicily.

A similar story had appeared earlier in an article on Union Sulfur by the *New York Times* (December 18, 1910):

> ...there stands the fact that the Union Sulfur Company, if it wanted to, could compete profitably with the Italian monopoly in all the markets in the world on such terms as would be ruinous to the Italian and would mean starvation to the island of Sicily.

However generous it may have been, the claim was condescending to the Sicilian sulfur industry, which had been relegated to a charity case. Unfortunately for the Sicilians, there was some truth in the statement.

The second five-year term of ASSC was coming to an end in July 1906. During the first term (1896 to 1901), with exports at record levels, Florio and the ASSC directors had been hailed for their business savvy. They had paid 50% per year on their common stock. Business accolades are a fickle beast; a hero one day, a goat the next. With the reversal in fortunes, praise switched to condemnation. The ASSC was blamed for the deteriorating situation, but there was little they could do except mitigate the damage. Under the circumstances, the ASSC had

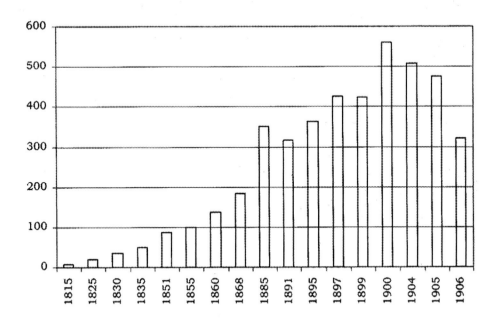

Figure 3.1. Sulfur - Sicily Exports from 1815 to 1906 ('000 tonnes).

negotiated an exceptional deal with the Americans. Their members did not see it that way. While dire warnings were given about the competitive position of Union Sulfur, most Sicilian producers refused to believe it. They would soon change their minds. The directors of the ASSC were frustrated by their members. In addition, personal tragedy had struck the Florio family when two of Ignazio's children died of typhoid in 1904; he was never the same afterwards. The directors had enough and decided not to renew their charter; the ASSC came to an end on July 31, 1906.

The Sicilian industry ruled the sulfur world for the entire 19th century. Despite periods of turbulence, this was a period of incredible growth, with exports peaking under ASSC in 1900 at over 550,000 tonnes (Figure 3.1). However, what had taken a century to create would be undone in but a few years.

An interesting theoretical exercise is what would have happened if pyrites had not taken over such a large share of the market in the second half of the 19th century. During the administration of ASSC, instead of selling 500,000 tonnes per year, their sales would have been well over one million tonnes per year, maybe even 1.5 million tonnes. More profits would have gone to the shareholders of ASSC, but would that have really changed the outcome? Frasch would still have destroyed the industry. One wonders, though, if the attitude of Herman Frasch would have been different. This astute entrepreneur would have realized that there was more opportunity than his one site at Sulfur Mine could handle. Frasch would have been more aggressive to develop new sulfur deposits. Perhaps Freeport Sulfur and Texas Gulf Sulfur would never have come to be; the only one would have been a giant Union Sulfur.

3.4.2 CONSORZIO OBLIGATORIO PER D'INDUSTRIAL SOLFIFERA SICILIANA (COISS): 1906 TO 1932

Whatever the short-comings of ASSC, the Sicilian sulfur industry was better off with them than without them. The members who had been critical of the directors of ASSC soon realized their folly. Making matters worse, as soon as the ASSC decided not to renew their term, they began ruthlessly dumping their stocks onto world markets, escalating the sulfur crisis. Between 1905 and 1906, five thousand miners lost their jobs in Sicily. The livelihood of over 250,000 people was in jeopardy. Riots broke out in Caltanisetta.

On July 15, 1906, the Italian government interceded forming a new monopoly, Consorzio Obligatorio per D'Industrial Solfifera Siciliana (The Compulsory Consortium for the Sicilian Sulfur Industry). This organization, with thirty directors, was one of the first government-formed trading cartels in modern history. COISS immediately stepped in to prevent the further dumping of the Sicilian inventories owned by ASSC. Sulfur was not to be sold below cost. To compensate for cash flow problems from sulfur in inventory, the government advanced 80% of the value of the material held in stock.

The first Director General of COISS, the distinguished mining engineer, L. Dompe of the ASSC, had been chosen by the original members. However, he resigned on June 29, 1907, and was succeeded by Pietro Lauro. His departure was likely associated with the control of COISS being taken over by a royal commission established by the government.

Fear remained within the industry. How powerful were the Americans? Could the previous tales be true? In early 1907, this time the Italian government sent a representative, L. Baldacci, to the U.S. to evaluate the Frasch competition. He only confirmed what Dompe had previously reported. The government report officially stated that Frasch was a serious threat and that the two groups should form an alliance. The report went on to state that U.S. sulfur could be delivered at less than $12 per tonne in Europe, while Sicilian production costs were $20 per tonne. The news was devastating to the sulfur industry of Sicily.

Making matters worse, since ASSC had folded, the deal with Union Sulfur was no longer valid. Upon its inception, the mandate of COISS was clear: make a deal with the Americans. Lauro and Frasch directly became involved. Frasch was less generous this time with his foreign competition. On February 29, 1908, COISS ceded one third of the European market to the upstart Americans (FSA-II).

In exchange, a global price of $22 per tonne (C.I.F. New York) was established. The deal collapsed again, by the passage of the "Five Sisters Act" in the U.S., which made illegal some of the terms of the agreement. Union Sulfur cancelled the agreement on January 20, 1913.

Frasch was no longer at the helm of the company, as he was ill and died in 1914. Union Sulfur started to flex its yellow muscles. In case COISS had any doubt about their competitive edge, Union Sulfur had ramped up production to an incredible 785,000 tonnes in 1912, greater than the entire global demand. A major marketing effort was launched in Europe and two new vessels were added to Union Sulfur's fleet. Sulfur was stocked at Rotterdam and Hamburg. COISS must have wondered if the Americans were willing to leave them with anything. Discussions began yet again between the two parties, but World War I broke out before any agreement could be reached. In 1914, Sicilian sulfur production was 303,886 tonnes, while total Italian output was 346,528 tonnes; U.S. production was almost 500,000 tonnes.

The War "protected" Europe from Frasch sulfur and global demand surged. Even so, this was a difficult year for the industry. Labor trouble and restrictions on fuel supply depressed production. In 1918, the term of COISS was renewed. After the war, the Sicilian industry was back on hard times. Trans-Atlantic vessel rates dropped, allowing the U.S. Frasch producers to be even more competitive in Europe. The two major markets, Britain and France, had switched mainly to American product. In 1921, the Italian government intervened to save the faltering business, a feature that the Sicilian sulfur industry would henceforth be dependent upon to survive.

COISS was still without an agreement with the Americans. Now, there were not one, but three companies to deal with: Union Sulfur, Freeport Sulfur and Texas Gulf Sulfur. In August 1922, COISS, led by the Royal Commissioner, Ernestino Santoro (~1887 - ?), hosted a "conference" in Geneva with Clarence Snider of Union Sulfur, Eric Swenson of Freeport Sulfur and Wilber Judson (~1880 - 1951) of Texas Gulf Sulfur. No agreement could be reached among the global competitors, nor even among the Americans themselves. Later meetings in New York, London and Paris also failed. At the Paris meeting, Donegani from Montecatini had been asked to attend. Montecatini had established its own cartel by purchasing more than 90% of the sulfur facilities in the Italian mainland in 1917. Most of the mines were soon closed, except for Perticara and Formignano (closed in 1962). He shrewdly told the Americans and COISS that their production of sulfur was so small, and even that was consumed internally, that his company should not be involved in the discussions. The other attendees agreed. Montecatini would later cause more problems for the already threatened Sicilian industry.

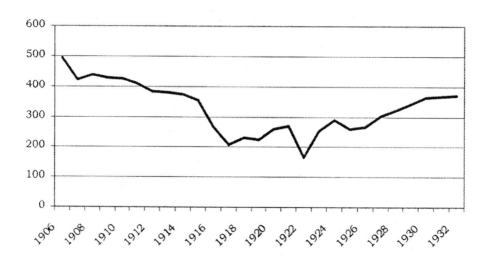

Figure 3.2. Sulfur - Italy. Production from 1906 to 1932 ('000 tonnes).

Finally, in a heated meeting in Rome, a deal (FSA-III) was reached between COISS and Sulexco (the new export consortium of U.S. producers) on March 14, 1923. The glory days of the ASSC only two decades before were now distant memories. The deal officially relegated COISS to a second-class global supplier. Sicily was guaranteed a minimum of 135,000 tonnes per year of sales. The Americans received 75% of the global markets, with the remainder to Sicily. These agreements were not clandestine; details were publicly available and filed with the FTC.

During the Depression, the fortunes of the Italian sulfur industry sank further. U.S. production was drastically cut back as global demand waned. Italian production, oblivious to the market reality, remained unchanged (Figure 3.2). Inventories grew at alarming rates. Sicilian producers were also facing new competition from within their own country; Montecatini was outside the control of COISS. The Sulexco agreement (FSA-III), though, applied to all Italian shipments. At the time of the signing, these only took place from Sicily. Now, 60,000 tonnes of exports were being shipped from Montecatini, which COISS had to compensate for. Thus, every tonne exported from mainland Italy resulted in one less tonne of exports from Sicily. Several attempts to get Montecatini to join the consortium failed, which proved to be the downfall of COISS. The latest Sicilian sulfur cartel was dissolved on July 31, 1932. The now familiar pattern continued. Following the dissolution of the latest cartel, the Sicilian industry, yet again, was in crisis. Some of the larger mines fell into bankruptcy. In 1932, total Italian sulfur production was 350,000 tonnes, of which 236,000 came from Sicily.

3.4.3 UFFICIO PER LA VENDITA DELLO ZOLFO ITALIANO (UVZI): 1934 TO 1940

On December 11, 1933, a new national consortium was formed called the Ufficio per la Vendita dello Zolfo Italiano (Central Sulfur Sales Bureau) in Rome, with C. Angelli as its first president. UVZI was controlled by six appointees by the Ministry of Corporations and the Ministry of Finance. The 200,000 tonnes inventory of COISS was first taken over by the Bank of Sicily and then by UVZI, to control supply and protect pricing. Their mandate was to sell the inventory over a six year period to prevent the repeat of the ASSC inventory dump. UVZI established a production and price quota system for all Italian producers, including Montecatini. The price quota was basically a price subsidy. Any shortfall between the fixed and world prices was to be reimbursed by a government fund, which turned out to be always the case. The government attempted to maintain production at 347,000 tonnes per year, of which 240,000 tonnes came from Sicily. Overall, UVZI was effective at maintaining the production quota and the price subsidy kept most of the mines alive.

A new organization meant that yet another new international agreement was required. UVZI and Sulexco met in London in July 1934. There was initial disagreement because UVZI was again insisting on global price increases to the government guaranteed price levels. One would have expected such cartel partners to willingly promote higher pricing, but not in this case. Sulexco knew that Spain and its pyrites, not Sicily, were their true competitive threat. As usual, Sulexco got its way and pricing remained the same. The agreement (FSA-IV; see Appendix I), though obviously favoring Sulexco, was fair to UVZI.

Despite the Depression, UVZI managed to hold the Italian industry together. In 1935, there were 106 mines operating in Sicily; most of which were small operations. Only seven were producing more than 10,000 tonnes per year; more than sixty mines produced less than 1,000 tonnes per year. Total Sicilian production was 225,000 tonnes. In contrast, Freeport Sulfur, alone, was producing this much sulfur from its two Frasch mines in Texas in only one quarter of the year! Sicily was also losing more market share to its own domestic rivals on the mainland. Between 1934 and 1940, Italian elemental sulfur production ranged from 300,000 to 350,000 tonnes per year (Figure 3.3), with almost half of this sulfur originating from the mainland: Sicily - 55%; Anacoria, Pesaro, Flori - 41%; Benvento - 4%. The start of World War II ended FSA-IV. After the war, further agreements were no longer warranted, as Sicily was no longer a factor on the global sulfur stage.

3.4.4 ENTE ZOLFI ITALIANI (EZI): 1940 TO 1962

With the start of World War II, control of the Italian sulfur industry switched from UVZI to Ente Zolfi Italiani (Italian Sulfur Board). Soon a joint German-Italian enterprise was formed called Societa per Incremento Produzione Zolfi, whose

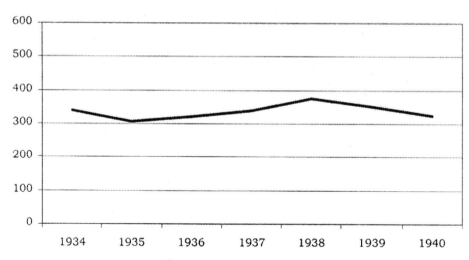

Figure 3.3. Sulfur - Italy. Production from 1934 to 1940 ('000 tonnes).

purpose was to improve the efficiency of the Italian sulfur industry. They were no more successful than earlier attempts. The number of mines in Sicily dropped to 74. In 1939 and 1940, Sicilian sulfur production was over 200,000 tonnes. By 1942, it was half this amount. Production virtually ceased the following year, when the allies invaded.

A little more than one hundred years after the Sulfur War of 1840, the British, along with American and Canadian troops, conquered Sicily and its aging sulfur mines. On July 10, 1943, Operation Husky (not the major Canadian recovered sulfur company!), under General Dwight D. Eisenhower (1890 - 1969), took place with the landing of 180,000 Allied troops near Syracuse. Sicily was defended by twelve Axis divisions under the leadership of General Alfredo Guzzoni (1877 - 1965). The island was taken within a month. The sulfur industry was down until December. Now under Allied control, sulfur production started slowly back up. In Sicily, 38,000 tonnes of sulfur were produced (a similar volume was produced in the mainland, still under Axis control) in 1944. The Allies established a new sulfur marketing firm, named Ente Zolfi Siciliani, and then two other organizations were formed, Unione Nazionale Zolfi in Rome, and Consorzio Zolfifero Siciliano in Palermo. The Sicilian industry, though, never recovered after the war, as Frasch and now the rapidly emerging recovered sulfur, eroded Sicilian market share even further.

The primitive and expensive Sicilian mines could no longer compete even with government subsidies. In 1945, U.S. sulfur was being delivered in Europe at $30 per tonne, while Sicilian sulfur was selling for 9,500 Lira per tonne, f.o.b. Sicilian port (exchange was varying; in 1945 at 100 Lira/U.S.$, increasing to 225

Lira/$ at the beginning of 1946). Even with the devalued currency, Italian sulfur was still not competitive. Sicilian mine production costs shot up to 30,000 to 40,000 Lira per tonne in 1948. While the government guaranteed price was raised to 28,000 Lira per tonne, actual selling prices were only 23,000 Lira per tonne. American sulfur was selling at an equivalent 18,000 Lira per tonne, C.I.F. Europe.

By 1948, Sicilian exports were only 60,000 tonnes, most of which went to France. The Sicilian sulfur industry was in a sorry state. Inventories had risen to 110,000 tonnes. The following year, the guaranteed (i.e., subsidized) price increased again to 34,000 Lira per tonne, as production costs continued to rise. Whereas EZI was paying the mines $55 to $60 per tonne in 1949, Frasch sulfur was selling for $18 per tonne, f.o.b. mine. The delivered cost of sulfur in Europe was much less from the U.S. than Italy. Sicilian sulfur was only going to areas not coveted by the Frasch producers. The only benefit of buying from Italy was the consumer did not have to pay in U.S. dollars, which were in short supply after the War. Within the Italian industry, the Sicilian mines suffered the most, because the continental Italian mines were better located to supply domestic needs.

A global sulfur shortage during the Korean War caused U.S. exports to be curtailed. This provided a brief reprieve to EZI, culminating with a special shipment of 150,000 tonnes from Sicily to Australia and New Zealand, at a price triple that of U.S. Frasch producers. This was the last hurrah of the Sicilian sulfur industry. Overall, exports had jumped to over 200,000 tonnes. During the sulfur crisis, the price jumped to 60,000 Lira per tonne. When the shortage subsided, Sicilian exports plummeted to below 10,000 tonnes in 1953. The industry was finished. The government had to issue emergency financial aid to the crumbling industry, providing EZI with a one billion lira loan. Another 900 million lira in subsidies were provided over 1957 and 1958. Between 1950 and 1962, the Italian government poured almost $775 million (U.S.) to support the struggling Sicilian sulfur industry.

In 1960, the European Economic Community (EEC) had approved tariffs to protect the terminally-ill industry and, the following year, established a special committee to see what else could be done. On April 24, 1969, the EEC approved of the Italian government's aid to the "unprofitable sulfur industry."

There was no longer any effort to try to modernize the industry; there was no point. The Sicilian sulfur industry was now only a government social program. Sicily could no longer afford to export any sulfur. The high production costs restricted Sicilian sulfur to mainly domestic usage. Between 1956 to 1962, the number of operating mines declined to 50, and total employment had been cut in half.

3.4.5 ENTE MINERARIO SICILIANO (EMS): 1962 TO 1985

While EZI continued to officially exist until mid-1968, control of the Sicilian sulfur industry had been given to Ente Minerario Siciliano (Sicilian Mines Authority) on December 22, 1962, when the Italian government passed the Minerals Law to deal with the faltering situation of the Sicilian sulfur industry. Their role was to consolidate the industry and relocate the displaced workers. The sulfur industry was to be wound down. In an especially humiliating position, Italy could not even afford to produce enough sulfur to cover its own demand. In 1963, sulfur was imported into Italy from Canada. The following year, EZI purchased 13,000 tonnes of sulfur from Sulexco. A few years later, against strong resistance by local politicians and labor leaders, the controlled phase out of the industry began as the number of mines declined:

1967 - 24 mines
1968 - 18 mines
1969 - 12 mines
1975 - 5 mines

Especially considering the tumultuous past, the process proceeded without major incident. By 1971, despite the consolidation, the production costs of the down-sized Sicilian sulfur industry were equivalent to $120 per tonne, while the

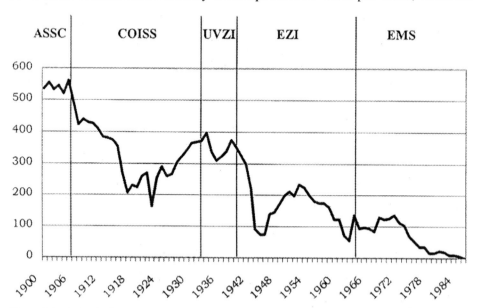

Figure 3.4. Native sulfur - Italy. Production from 1900 to 1985 ('000 tonnes).

global price was $38 per tonne. The few remaining mines were each losing $4 million per year. In 1972, production was only 115,000 tonnes. Among the last operating mines were Giangagliano, Floristella-Grottacalda, and Giumentaro. None of the mines could be saved, as even the best of them were non-competitive. Where over one hundred mines had existed fifty years before, the last native sulfur mine in Sicily closed in 1985. As with most industries, the Sicilian sulfur industry had ended with a whimper (see Figure 3.4). Today, Italy is still a major sulfur producer in Europe, but only recovered sulfur from oil refineries.

FRASCH SULFUR — TEXAS/LOUISIANA

In Texas and Louisiana lay the world's largest reserves of elemental sulfur. In central Florida lay the world's largest reserves of phosphate rock. The fortuitous proximity of the world's largest sulfur supply and the world's largest sulfur consumer produced the greatest sulfur companies in the history of the industry.

4.1 NATIVE (NON-FRASCH) SULFUR MINES

The U.S. was already one of the few sulfur producing countries in the world by the mid-19th century. In the first half of the 20th century, small native-sulfur mines operated in the U.S., usually unsuccessfully, in Utah, Texas, Colorado, Wyoming and Nevada; the most important sulfur mining region was Inyo, CA. Two of the more famous early U.S. sulfur mines were the Rabbit Hole Mine, operated by the Nevada Sulfur Mining Company, and the Dickert and Myers Mining Company in Utah. Production of mined sulfur in the U.S. only amounted to a few thousand tonnes per year until the mid-1950's. Native sulfur production surged to 87,000 tonnes in 1957. By the mid-1960's, the native mines had all but disappeared.

Besides the small mines in the West, one large sulfur deposit had been found at Calcasieu, LA. However, unlike the surface native sulfur deposits, this one was found deep in the ground. In 1867, the Louisiana Petroleum and Coal Oil Company (LPCO) had first located the sulfur cache, near Lake Charles, LA. Two years later, the geologist Professor Eugene Hilgard (1833 - 1916) from the University of Mississippi was invited by the company to inspect the interesting geological deposit. This visionary immediately saw the Louisiana sulfur as a competitive supply to Sicilian sulfur. Although he believed that it would be a challenge, the German-born professor recommended to go ahead with mining operations. The unique American sulfur deposit became known to the world when Hilgard published a paper on his investigations in the *Engineering and Mining Journal*, in September 1869. [A few years later, in 1873, Hilgard left Mississippi to become the first professor and, later, director of the College of Agriculture at Berkeley. By 1880, he had started a scientific study of wine making. His pioneering work helped establish the California wine industry. His memory is honored by Hilgard Hall, which houses the Soils Department today at Berkeley.]

When LPCO attempted to switch their operations from oil exploration to sulfur, they found that their lease agreement was only for oil, and they lost the sulfur rights. In many respects, they were lucky that they had. The world was not yet ready to tackle this technological and physical quagmire.

In 1870, the Calcasieu Sulfur Mining Company (CSM) of New Orleans was the first to fail at extracting the gold of the bayou. The general manager of CSM was Captain Ino A. Grant. The company brought in the noted French mining engineer, Antoine Granet; his geological study was the first to scientifically describe the structure of the salt dome with its sulfur and limestone cap. However, even the leading mining experts of the day could not overcome the quicksand that lay above the sulfur. The search for this yellow buried treasure had much of the same challenges as the mysterious treasure of the Oak Island Money Pit. For nine years, CSM tried to reach the sulfur deposit, but never came close.

In 1879, they sold their rights and assets to the Louisiana Sulfur Company (LSC), under the direction of Duncan F. Kramer (according to Haynes, or Kenner according to Sutton). Their attempts were no more successful. LSC kept control of the property for the next ten years. They, and others who sub-leased from them, for example the National Sulfur Company in 1886, also failed. In July 1902, the National Sulfur Company, led by Charles Dobson, tried again. They were going to overcome the quicksand by freezing the ground with ammonia, and then dig their mine. By this time, Union Sulfur was operating, so they looked just outside of their rights. However, they never found any sulfur to test their technology on.

Twenty years had passed since CSM had first dug into the quicksand of Calcasieu, and no one had yet come even close to the sulfur trove. In 1890, LSC sold their rights to the American Sulfur Company (ASC). This elite investment group included Abram Stevens Hewitt (1822 - 1903); his son Peter Cooper Hewitt (1861 - 1921); his brother-in-law Edward Cooper (? - 1905); Hamilton McKown Twombly (1849 - 1910), friend and financial advisor to Hewitt; William T. Sanger; Richard P. Rothwell (1836 - 1901); William Russell Grace (1832 - 1904); Alexander Harvey Tiers (~1857 - 1910) and his brother Cornelius Tiers (~1857 - 1921), who had long worked with Hewitt and Cooper. Among this prestigious group were three mayors of NYC from the 1880's, Cooper in 1879 and 1880; Grace in 1881, 1882, 1885, 1886; and Abram Hewitt in 1887 and 1888. Abram Hewitt, also a five-time congressman, had been a founder of the U.S. Geological Survey and had been a president of the American Institute of Mining Engineers. His son Peter had a Ph.D. from Columbia and had invented the mercury-arc lamp. W.R. Grace is best known for the chemical company he founded that was named after him. The millionaire Twombly was a successful investor. He had married Florence Adele Vanderbilt (1854 - 1952), granddaughter of the legendary Cornelius Vanderbilt (1794 - 1877), in 1877. Rothwell was a well-known mining engineer and later became editor of *Engineering & Mining Journal*. The assembly of such intellectual and financial capital into one syndicate was still not enough to overcome the challenge of the sulfur in the bayou.

The first president of ASC was Edward Cooper, until 1893, when Alexander Tiers took on this role. ASC raised $1 million in capital to finance the project. They restarted the project proposed by Granet, whereby iron rings were used to keep back the quicksand. The company aggressively pushed deeper towards the elusive goal. Granet and CSM had never got past 110 feet. This time the mine reached 250 feet but again quicksand stopped them from going any further. In November 1893, ASC drilled its last well, the thirteenth such well since the days of CSM. Unlucky #13 proved to be worse than any before them, as five workers died from a massive hydrogen sulfide gas release. The enticing, but deadly, sulfur deposit was beyond the reach of the technology of the day. Just as another venture was about to fail at Calcasieu, the man with the magical technology walked in the door. His name was Herman Frasch.

4.2 HERMAN FRASCH

4.2.1 HIS LIFE & FAMILY

On December 25, 1851, Herman Frasch was born in Gaildorf, Germany, into a middle-class family. The small town, with a population of about 1,600, is in the state of Baden-Württemberg, to the east of Stuttgart. His father, Johannes (1811 - 1891), was burgomaster of Gaildorf from 1854 to 1866. His mother was Katherine Baur (1819 - 1891). Herman, the eldest son of the family, was born late in the life of his father, who was already forty years old when Herman was born. Starting a family at such a late age was uncommon in the mid-19th century, unless Johannes had been married previously (there is no record of this). Herman attended a local elementary Latin school and was then an apprentice at a local pharmacy, a common route to gain practical knowledge of chemistry in these still early years of this science. His father then sent the blue-eyed, red-haired lad to Schwabisch-Hall (population 7,000), ten miles to the north of Gaildorf, to apprentice in a book shop. Hall (meaning "salt") was the site of an ancient salt mine that was still an important industry in the region. Little did the teenager know that the mining of salt would later play a role in the development of his fame. His stay at Hall was interrupted after a prank by the spirited Herman. He made the town clock strike twelve times in the middle of the night. The residents were not amused and neither was his boss; he was sent packing back to Gaildorf.

After his untimely prank, the young Herman told his family of his desire to go to America. His father, though, sent him to the gymnasium at Halle (by the Royal Frederick University; now known as the Martin Luther University), in Saxony, another city famous for its ancient salt works. After a short time at Halle, Herman returned to Gaildorf again prodding his father to let him go to America.

His parents agreed to let the sixteen year old ship off alone to the U.S., where an uncle, George Frasch (~1827 - ?) was living in Philadelphia, after emigrating in 1854.

Herman arrived in Philadelphia at the home of his uncle in August 1868. As discussed previously, Philadelphia was the heart of the early American chemical industry and was the center of sulfuric acid production. The young Frasch, though, did not join one of the major chemical firms, but worked in the pharmacy business. Then, pharmacies were more than dispensers of medicines. They were a place to obtain practical training in chemistry outside of university and provided chemical consulting services.

In 1875, Herman's younger brother, Hans A. Frasch (1856 - ?) also moved to the U.S. On July 26th, he arrived at New York City on board the *Nevada*. Ironically, he then moved to Louisiana, where Herman would later gain his greatest fame. The brother of Herman was also a chemist, first working in a salt mine and then in the sugar refining industry. In the 1880 U.S. Census, his occupation was listed as a druggist, and he was living at the home of Estella Hardy. While in Louisiana, Hans married Bertha L. (~1861 - ?); Bertha died between 1910 and 1920. Herman found his brother a position at his firm the American Chemical Company at Bay City, MI. Hans was also an inventor, but not as proficient as his older brother. One of his early discoveries was on oil lamps. In 1886, he had entered into a partnership with a group in Pittsburgh to develop the invention. Also, Herman transferred to Hans his own patent (U.S. Patent 340,711) on an oil lamp design that he had filed on October 14, 1885. In 1887, Hans moved to Austin, TX, after the closure of the American Chemical Company, where he worked for Graham & Andrews. He also briefly worked for the Grasselli Chemical Company. When Herman opened his second alkali venture, Frasch Process Soda Company, Hans returned to Cleveland to work at the facility. He later moved to New York City in 1898, establishing a private, consulting office. Hans later made several discoveries in the petroleum industry, specializing in dye manufacturing.

Upon arriving in America, Herman wasted no time to develop his career or to find romance. In 1869, he married Romalda Berkin (1854 - 1889; her last name may also be Berkie, Berkeley or Burks), who was born in Pennsylvania. Her mother was also from Pennsylvania, while her father was from Hesse, Germany. The bride was only fifteen years old, and Herman seventeen years old when they got married. The young couple continued to live with his uncle George, until they moved to a flat above Taylor's pharmacy, where Herman had found a job, on Race Street and 9th. Herman's career saw the family move several times, first to Cleveland in 1877 when he joined MMPC. In November 1882, West Bay City, MI, was their next home, as Frasch operated his American Chemical Company plant here, and then they moved to London, ON, in May 1884, when he was consulting for

Imperial Oil. The family was off again, back to Cleveland in 1888/9, when he became an executive with Standard Oil. His young wife Romalda died about this time. There was no record of her death in newspapers or city records in Cleveland; she may have died in London, ON. Herman was left (briefly) to raise two teenagers, George and Frieda, by himself.

Their eldest child George Berkin Frasch, had been born in Philadelphia, on October 27, 1873. George, named after his uncle, became a chemist in Cleveland. He had taken some interest in his father's sulfur business, visiting the site between 1897 and 1903. In September 1906, he was to have married Nancy Lovis of Cleveland. The wedding was called off at the last moment, allegedly because of the objections of the bride's mother. Shortly afterwards, the despondent George then immigrated to Sidney, Australia. George married Ruby Graham (1879 - 1954) on April 10, 1907, in Australia. They returned to the U.S. in 1909. In 1920, George was working as a chemist in Los Angeles. Herman and Romalda had a second son, Herman Frasch Jr., who was born on October 6, 1875, but died in infancy.

After moving to Cleveland, Herman and Romalda had a daughter Frieda (1879 - 1951) in August 1879. On October 15, 1902, Frieda married Henry Devereux Whiton (1871 - 1930), son of Edward Nathan Whiton and Mary Devereux. The families had been neighbors on Euclid Avenue in Cleveland, the Frasch's at 890 and Henry was living with his grandmother, Antoinette C. Devereux (~1832 - 1911), at 882. The Frasch mansion on Euclid had 18 rooms and 5 baths. The family was living in Hempstead, New York, by 1910. Frasch made his son-in-law, vice-president of Union Sulfur. Henry later became president of the firm upon the death of Herman in 1914. During World War I, he headed the Sulfur Committee of the War Industries Board. Their son Herman Frasch ("Swede") Whiton (1904 - 1967) was born on April 6, 1904. In 1926, he joined Union Sulfur after graduating from Princeton. The grandson of Herman became president (the third member of the family to hold this position), after George Miller Wells (~1880 - 1957). In 1952, he retired and was replaced by Richard T. Lyons (~1896 - 1979). Herman Frasch Whiton was a noted yachtsman, specializing in the six-meter class. He was a member of the U.S. Olympic team, first at Amsterdam in 1928 (sailing the *Frieda*; finishing sixth), and later winning a gold medal at the London Olympics in 1948, and then repeating the feat at the Helsinki Olympics in 1952. He was the first American to win back-to-back gold medals. Herman Frasch Whiton died on September 14, 1967. He had married Emelyn Thatcher Leonard (~1916 - 1962) on June 28, 1939. Their children were Emelyn Patterson, Romalda Berkeley Clark, Charles Whiton and Herman Frasch Whiton Junior. The couple divorced in 1957. The following year, he married Kathleen O'Brien (1916 - 1997).

Frieda and Henry Whiton were divorced in 1921. On September 18, 1922, she married Count David Augustus Costantini (1875 - 1936). The Count was a diplomat, who had served as a counselor at the Italian embassy in Washington. He and Frieda spent most of their time in Europe in the social circles of the aristocratic elite. Not long after the death of the Count, on June 23, 1937, Frieda remarried again, to Baron Carl Gottlieb von Seidlitz (1872 - 1954). At their wedding in Jacksonville, FL, Herman Frasch Whiton gave his mother away. Frieda remained married to the Baron until her death on January 8, 1951.

Not long after the death of Romalda, on June 16, 1890, Herman Frasch married Elizabeth Blee (1858 - 1924). Her parents were James Blee (~1830 - ?) and Elizabeth Blee (~1840 - ?), who had been born in Canada; James' brother, Robert Blee (~1834 - 1898), became mayor of Cleveland in 1893. For their honeymoon, Herman and Elizabeth went to Europe, including a visit to his parents. Frasch was fortunate to have seen them one last time, for both of them died the following year. Afterwards, he routinely returned to Gaildorf to visit their graves. Frasch made frequent trips to France and Germany, going there almost every year from 1903 until he moved to France shortly before his death. In 1908, the now famous Frasch was honored by his hometown being declared a "Freeman of Gaildorf." At this occasion, he had donated a festival hall to the city (the city landmark was destroyed in 1945). This car enthusiast had arrived in a convertible that was yellow…of course, representing sulfur! Besides his many ventures, Herman's favorite hobby was his cars; he had been a member of the New Orleans Gentlemen's Driving Club. Four years later, Frasch was in Berlin and again brought an automobile. At 135 h.p., such a powerful machine had seldom been seen in Berlin streets before. His love for cars got him into trouble, during his 1912 vacation. Herman met Robert Loonen on his way to Europe. Loonen convinced Frasch to loan him $35,000 to float a new automobile company. Loonen, though, was a crook and swindler.

When Elizabeth Frasch died on September 25, 1924, in Paris, she had left instructions that she be buried in the mausoleum at Gaildorf beside her husband. However, after her death, Herman was reburied in North Tarrytown, NY (now called Sleepy Hollow), near New York City, at the Sleepy Hollow Cemetery. The Frasch Mausoleum was then donated by the family to the town of Gaildorf, where it remains a tourist attraction. Elizabeth was buried at Locust Valley, on Long Island, at the estate of her step-grandson Herman Frasch Whiton. Her will was contested by members of the family. The bulk of her one-million-dollar estate had been left for "research in the field of agricultural chemistry, with the hope of attaining results which shall be of practical benefit to the agricultural development of the United States." The fund became known as the Herman Frasch Foundation for Chemical Research. After several court battles, the will was finally upheld by the Court of Appeals in July 1927. The first funding was not made until early 1929. The Herman Frasch Foundation continues to exist today under the administration of the American Chemical Society.

Herman Frasch's career spanned forty years. After a brief apprenticeship, his first job in Philadelphia was at a pharmacy owned by William Taylor. The following year, he joined Professor John Michael Maisch (1831 - 1893) of the Philadelphia College of Pharmacy, who later became a founder of the American Chemical Society and the editor of the prestigious *American Journal of Pharmacy.* In 1873, Herman left Maisch's pharmacy to work for another firm. At the end of the year, the independently-minded German entrepreneur established his own business in Philadelphia. His intellectual and creative energies could never be contained within one venue for long. He was always searching for the next great challenge to overcome. Frasch, though constantly moving from one venture to the next, was never superficial in his studies. Quite the opposite, he was detailed to a fault. The following year, he formed a partnership with another pharmacist, the German John M. Ruegenberg (~1843 - ?), and together they formed the Philadelphia Technical Laboratory. The company's specialty was brewing and wines. Frasch's interests had started to diversify. A local machinist asked what could be done with tin scraps. A business venture evolved, but it quickly failed. Later, his first successful patent was filed on tin scraps (U.S. patent 171,276), on July 24, 1875. Herman formed a partnership with Charles B. Sprogell (~1838 - ?) and Henry Thomas to exploit the technology. While these inventions were for the technical problems of others, he was looking for business opportunities incorporating his technologies for himself. Never an academic, Herman Frasch was a pioneer in the emerging area of applied chemistry. He was among the first and among the best! His creative genius combined with a practical sense opened industry's eyes to the value of the chemist.

Frasch's career extended across several fields, but the majority of his work was associated with oil and, later, sulfur. Frasch was an entrepreneur extraordinaire. He was a compulsive creator of new ventures, all creative, but seldom successful. The one great success more than made up for the failures. The business enterprises had started as a young lad in Philadelphia. His first major industrial employer was MMPC in Cleveland (1877 to 1884), a major paraffin wax manufacturer that was controlled by Standard Oil. His entrepreneurial spirit exploded during the 1880's and 1890's with the creation of many ventures: the American Chemical Company (1881 to 1888); Empire Oil (1885 to 1888); Spence, Frasch and Company (~1885); Frasch Process Soda Company (1890 to 1905); United Salt Company (1892 to 1899) and Union Sulfur (1890 to 1914). While all this was going on, Frasch continued to explore other areas. The years 1894/1895 especially exhibit the drive and diversity of this compulsive entrepreneur. In 1894, he developed a new method of purifying linseed oil (U.S. Patent 550,716) for the Cleveland Linseed Company. At the same time, he found the time to study gold mining (U.S. Patent 565,342) and had suggested a venture to John D. Rockefeller. At the end of the year, he was down in Louisiana conducting the first trials of his revolutionary sulfur technology. Early the following year he was working on enhanced oil recovery with Van Dyke; the technology was assigned to the company of the latter, the Oil Well Acid Treatment Company of Lima, Ohio. At this same time, he was already involved with his Frasch Process Soda Company and the United Salt Company. Remarkably through most of this period, he was also a "full-time" employee of Standard Oil (1886 to 1899)! He became involved in so many independent projects that one suspects that Frasch was the perpetual entrepreneur.

As his later sulfur business began to finally flourish, Herman spent more and more time at the company's head office in New York City, first at 12 Broadway and then moving to 82 Beaver Street in 1904. Frasch usually stayed at the Waldorf-Astoria while in NYC. While keeping his home in Cleveland, he also purchased a house in Paris, on the Avenue de Malakoff, near the Place d'Etoile and a vacation home at the trendy Saint-Jean Cap Ferrat on the French Riviera between Nice and Monaco. From 1911, he listed his official residence as being in France.

For his contributions to applied chemistry, especially sulfur and oil, he was honored by the Royal Society with the prestigious Perkin Medal (1912). The award was conferred on December 19, 1911, only six days before his 60th birthday, and the presentation ceremony took place on January 19th, at the Chemists' Club in New York City. A fragile Frasch attended the ceremony and summarized his career. While he proudly reviewed his technology to process the Ohio crude into a high-value product, most of his speech focused on his beloved sulfur process. Other presentations were given by the past president of the Society of the Chemical Industry and a founder of the American Chemical Society, Charles Chandler (1837 - 1925), the flamboyant Captain Lucas, who had discovered the gusher at Spindletop and had dug into every salt dome he could find in the Gulf afterwards, and Francis H. Pough (~1869 - ?), who had been a Union Sulfur executive since 1898. Plough gave an overview of the operations at Sulfur Mine, the plant that Frasch built. Union Sulfur went all out for the presentation to honor their founder and president. The presentation included thirty lantern slides and a silent motion-picture of the sulfur mining operations!

After the medal presentation, Frasch went to Pasadena, CA. His son George and wife Ruby were living in this area. Perhaps, he visited his son, whom he had not seen since he had left for Australia.

Frasch had been suffering for many years from Bright's disease (glomerulonephritis, an inflammation of the kidneys), which kept him at spas and health resorts in Europe. He was forced to retire because of his declining health. After a brief return to New York in April, he left for Paris in June 1912. Besides a brief visit to New York City and to Gaildorf in the summer of 1913, he stayed in Paris.

One of his last commercial endeavors was negotiations with Swenson and Vanderlip of Freeport Sulfur for rights to the use of his technology at their new sulfur mine in Texas. The negotiations were never completed. In his last tribute, Frasch was made a director of the prestigious Research Corporation in 1914. Frasch did not serve long, for he died in Paris on May 1, 1914. Frasch was first buried in his hometown of Gaildorf. His wife and daughter had built a large mausoleum for him there. On June 9, 1914, Elizabeth and Frieda had arrived in New York City on board the *George Washington*, sailing back from Germany, after burying Herman.

Most of his estate was divided between Elizabeth and Frieda. George Frasch was provided with only a trust fund of $50,000. The estate was first declared to be worth $5 million. However, William W. Wingate (~1871 - 1939) of the State Controller's Office contested the evaluation on March 9, 1915, claiming that it should be more than $25 million. If so, the state was due more inheritance taxes. The dispute was over the value of the stocks of Union Sulfur. A decision was reached in October 1916, and the value of the Union Sulfur shares was raised from $1.1 million to $4.9 million. The change made the overall estate worth more than $10 million. The debacle over the shares demonstrated the value that Frasch had created in Union Sulfur (*New York Times*, p. 22; October 2, 1916):

> The increase in value of the Union Sulfur stock from practically nothing to $12,003 a share is due almost entirely to Mr. Frasch's inventive mind and his ability to solve problems believed to be unconquerable.

A lasting memorial to Herman Frasch is found in Sulfur, LA, in Calcasieu. The town had come into being because of the nearby sulfur mine. The earliest settlers of the region had arrived during the days of CSM. The first building, a general store, had been built by Elias A. Perkins (1833 - 1917) in 1876 (closed in 1891). John Thomas Henning (1842 - ?), the husband of Eli's daughter Catherine (1860 - ?), built the first house in town in 1885. Their son was John L. Henning (1880 - ?), who worked most of his life for Union Sulfur, finishing his career as vice president and general manager at the mine. In 1910, Frasch gave a donation to the town to build the first brick school house, Frasch School. The Frasch Elementary School still exists in Sulfur today, at 540 South Huntington Street. When Frasch died a eulogy was given at Sulfur, by Congressman Arsene Paulin Pujo (1861 - 1939). He was congressman of the region from 1903 to 1913 and had acted as legal counsel for Union Sulfur. Shortly afterwards, on June 16th, the site with a population of 1,700, was officially designated a town; Dosite Samuel Perkins (1866 - ?), son of Eli, became the first mayor. The town became a city on January 11, 1952, and now has a population of 22,000. Sulfur is the home of Frasch Park, which includes Frasch Park Gymnasium, the North Frasch Softball Complex and Frasch Park Golf Course. Not far from the park is the Brimstone Museum, opened in 1976 to honor Frasch and his pioneering sulfur company, Union Sulfur.

4.2.2 HIS OIL VENTURES

In his early career in Philadelphia, Frasch started looking at petroleum. At the time, the industry was still in its infancy. Oil had been discovered in Western Pennsylvania in 1859, and markets for its products were only just being developed. The dominant market was kerosene for oil lamps, followed by lubricating oils and greases. Kerosene had been discovered by the Canadian geologist Abraham Pineo Gesner (1797 - 1864). In 1850, he had formed the Kerosene Gaslight Company in Halifax, NS, and four years later, the North American Kerosene Gas Light Company in Long Island, NY (later taken over by Standard Oil). Gesner first isolated the product from coal ("coal oil") and later developed the process for isolating it from petroleum. Other petroleum products that had entered most pharmacy shops, where Frasch was working, were Vaseline and paraffin wax (mostly used for candle manufacturing). In late 1876, Frasch discovered a new method to purify paraffin wax (U.S. patent 190,483; Can. patent 7,691).

The initial work of the unknown Frasch was so impressive that one of the largest wax companies made him a job offer. Joseph B. Miriam (~1827 - ?) who was vice-president and general manager of the Cleveland firm Meriam & Morgan Paraffine Company (MMPC), a major manufacturer of paraffin wax candles, and greases and lubricants, contacted Herman. The company had been founded in 1863 as Morehouse and Morgan. Meriam's father-in-law, Edward P. Morgan, was president of the company, and William Morgan was superintendent. In 1873, Standard Oil (ExxonMobil), a major supplier to MMPC, had bought into the firm. By 1880, the company was employing 72 people. In 1877, MMPC offered Frasch a job, and he moved to Cleveland, the center of the early American petroleum industry. He worked at the MMPC plant, by the Standard Oil refinery. After the kerosene was removed, the heavy oil remained, where was found the paraffin and lubricants. Frasch wasted no time in this new position, filing his second petroleum patent, an improvement on his first, on August 3, 1877 (U.S. patent 205,792). Another was filed on December 6, 1879 (U.S. patent 231,420). Frasch was extracting commercial products from what otherwise was a waste to the petroleum industry of the day.

Frasch's work continued to impress. His work on heavy oil came to the attention of the new major petroleum company north of the border, Imperial Oil. On September 8, 1880, sixteen of the oil refineries in the Sarnia area had merged together to form Imperial Oil; Frederick A. Fitzgerald (~1840 - ?) was the company's first president. One of the founders and first vice-president, Jacob L. Englehart (1847 - 1921) started negotiations with MMPC in the fall of 1882, for the rights to Frasch's petroleum patents, and an agreement was reached in January 1883. The price was high for the technology of the young German. Frasch and MMPC were paid $10,000, and Meriam was given 500 shares of Imperial Oil stock worth $50,000, and another 500 shares were placed in trust for Frasch, Meriam and Morgan. Frasch and Meriam were also elected to the board of directors of Imperial Oil. However, Imperial was very pleased with the outcome of their hefty investment. Imperial obtained the rights to the patents (Can. Patent 19,189 was assigned to Imperial) and to the use of Frasch as a consultant to get the technology up and running at their refinery.

Shortly after the deal had been concluded, disaster struck both companies. In February 1883, the Cuyahoga River flooded. The main Standard Oil refinery was saved, but some oil tanks were toppled and the oil spilled into the river. A massive fire took place, and the MMPC facility had to be rebuilt. On July 11th, lightning struck the main Imperial Oil refinery in London, ON, and the facility was destroyed. Operations were consolidated at the company's other major refinery at Petrolia, ON.

Frasch was still living in Bay City throughout 1883, spending most of his time on the alkali works of the American Chemical Company for MMPC (see below). On May 17, 1884, his last patent (U.S. Patent 304,309) for MMPC was filed on the manufacture of wax paper. This led to a brief joint venture with William Melville Spencer (1852 - 1931), of the Victor Oil Works and a director of Imperial Oil, who later became mayor of London in 1892. Spencer, Frasch and Company soon folded and the rights to the technology in Canada (Can. Patent 22,663) were sold to James Harvey McNairn of Toronto, who had entered the wax paper business in 1882 (McNairn Packaging still operates today in Whitby, ON). The U.S. rights to the technology were eventually sold in 1887 to the Hammerschlag Manufacturing Company of New York, owned by Siegfried Hammerschlag.

In May 1884, Frasch moved to London, ON, and went to work installing his technology for Imperial Oil. His salary was the same as the president of the Canadian oil giant. Reports on the early Canadian industry mention Frasch working for Imperial Oil as its "First Chemist." The relationship between Frasch and Imperial is a bit exaggerated, since he only officially worked for them for less than a year. After completing the technology transfer, he resigned from his lucrative position in February 1885.

After leaving Imperial Oil, Frasch and John R. Minhinnick (~1838 - ?), another director of Imperial Oil, purchased an idle refinery in Petrolia, ON, and formed the Empire Oil Company of London, ON. While in London, Frasch continued his studies on petroleum. On October 14, 1885, a patent (U.S. Patent 340,711) was filed on a new oil lamp, but the ownership remained with Frasch and not Empire Oil (later transferred to his brother Hans). Another patent (U.S. Patent 340,999) on oil refining was filed on November 11, 1885. Frasch was also back working at alkali manufacturing; a series of patents were filed in early May 1886 (see below). Again, Frasch retained control of the patents. Soon, though, more pressing issues were taking up the time of this creative genius.

Frasch's first serious claim to fame took place in Canada. The fledgling Canadian petroleum industry in Ontario, which was producing more than one million barrels per year, was in trouble because their oil was sour (i.e., it was contaminated with sulfur). Imperial Oil had problems selling the product because of its foul odor and the soot it produced when burnt. The Ontario crude had become known as "skunk oil." Now at Empire Oil, Frasch discovered that copper oxide economically removed the sulfur from the oil (forming copper pyrites), and could be easily regenerated and used again. The famous pyrites-roasting MacDougall furnace was utilized by Frasch for this regeneration. Frasch had captured the "Canadian skunk." He built a pilot unit at the Empire refinery to treat 1,200 barrels per day of oil with his revolutionary process, and his company was selling this refined Canadian oil. In 1887, Empire Oil claimed that their illuminating oils (i.e., kerosene) did not "smoke or emit any disagreeable smell while burning; they are manufactured by a patented process." These trademark oils, presumably made by the Frasch process, were called "Royal Palace Light Illuminating Oil" and "Aurora." His breakthrough patent was filed on February 21, 1887 (U.S. patent 378,246; twenty related patents were applied for by the end of 1894). The Canadian patent was not filed until April 16th (Cdn. Patent 28,750); the first page of the patent is given in Figure 4.1.

This famous patent was the focus of the Empire Scandal. Frasch had a double identity (i.e., employer) while with Empire Oil. The first hint of this originates in the front page of the patent: "Refining Canadian and Similar Oils." The "similar oils" were those from Lima, OH, the giant oil field that had recently been taken over by Standard Oil. Frasch had dealings with Standard before, since they held controlling interest in MMPC. He had become reacquainted with his petroleum friends in Cleveland several months before the patents were even filed; he was, in fact, working for Standard Oil!

The sudden interest in Frasch by Standard Oil had been sparked by a chance discovery. In the spring of 1885, just after Frasch had joined Empire Oil, a huge oil find had taken place in Lima, OH.

However, most firms passed on the opportunity because the oil was sour, the same problem as in Ontario. John Davison Rockefeller (1839 - 1937) gambled on

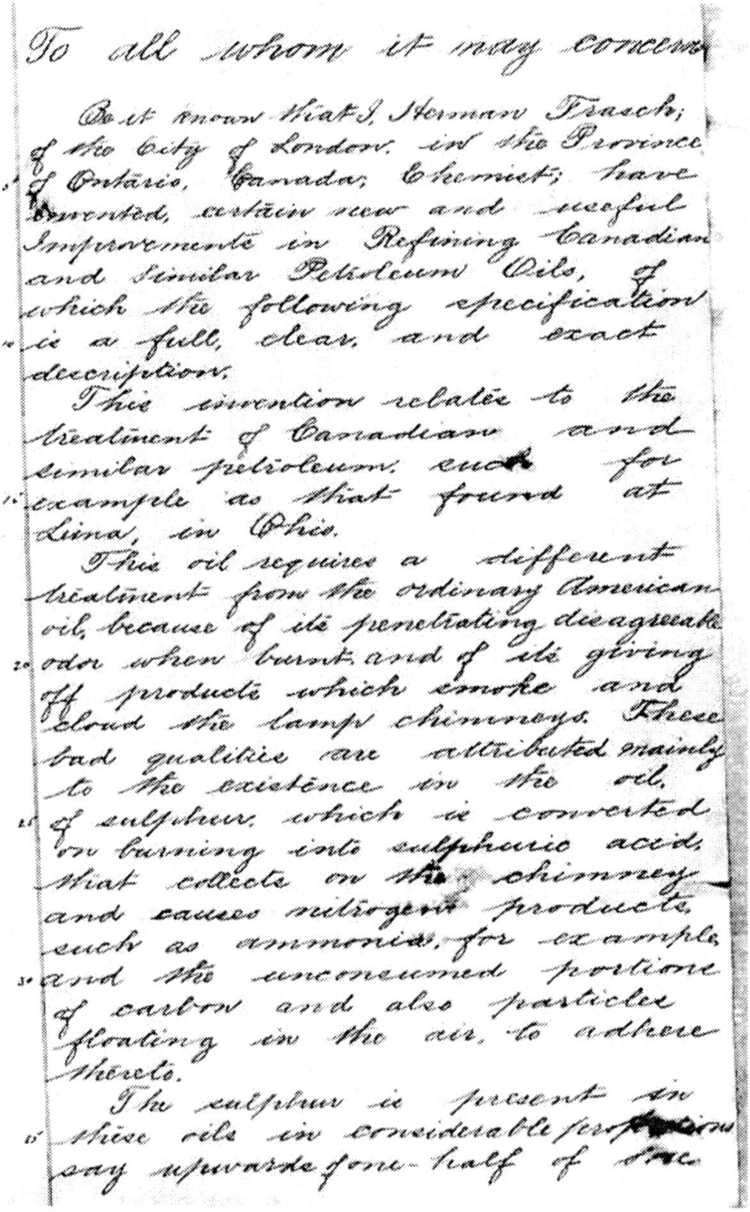

To all whom it may concern

Be it known that I, Herman Frasch, of the City of London, in the Province of Ontario, Canada, Chemist, have invented certain new and useful Improvements in Refining Canadian and Similar Petroleum Oils, of which the following specification is a full, clear, and exact description.

This invention relates to the treatment of Canadian and similar petroleum, such for example as that found at Lima, in Ohio.

This oil requires a different treatment from the ordinary American oil, because of its penetrating disagreeable odor when burnt and of its giving off products which smoke and cloud the lamp chimneys. These bad qualities are attributed mainly to the existence in the oil of sulphur, which is converted on burning into sulphuric acid, that collects on the chimney and causes nitrogenous products, such as ammonia, for example, and the unconsumed portions of carbon and also particles floating in the air, to adhere thereto.

The sulphur is present in these oils in considerable proportions say upwards of one-half of one

Figure 4.1. The first page of Canadian Patent 28,750.

the oil fields and Standard Oil moved quickly to gain control of the oil. Without the removal of the sulfur, the investment would be a disaster. The risky venture would not work without Frasch. Rockefeller gave Frasch an offer that he could not refuse, as the German entrepreneur sold his sour-oil soul to the Standard devil. On July 1, 1886, Standard Oil bought Frasch's yet unfinished and unpatented technology. Four days later, he was in Cleveland to brief the Standard Oil executives on the technology. The process worked, but many refinements were necessary to make it practical. He returned to London to continue his experiments at Empire Oil for Standard Oil. Meanwhile, Empire Oil had installed the Frasch technology, but they would soon be forced to close their plant, since the technology was now owned by their major competitor.

Rockefeller, not one to wait around, in December 1886, already commenced building a refinery for the sour crude, the Solar Refining Company, at Lima, OH, with Frank Rockefeller (1845 - 1917) as its president, even though the Frasch technology was still under development. The refinery was ready before Frasch was. The end result was that most of the initial production from the Lima field could not be sold but had to be stored. At the same time, the huge influx of low-grade oil sent prices plummeting. Standard Oil did sell some Ohio oil but only as a fuel at costs less than coal ($0.14 per barrel), which was nowhere near the premium of the "sweet" Pennsylvania oil ($2.25 per barrel). The only major difference between the two was less than 1% sulfur in the former. During 1887, Rockefeller's daring gamble seemed to be doomed to fail.

At last quitting Empire Oil, Frasch moved back to Cleveland in 1888 and became the first director of R&D for Standard Oil, working directly for Rockefeller. During the first half of 1888, the Lima oil was still a financial disaster, as Frasch tirelessly worked to get his process up to the scale. Besides most of the sales being at depressed pricing, inventories had risen to almost ten million barrels. Rockefeller's partners were urging him to write-off the investment. In September 1888, Frasch was ready, and the new technology was immediately installed at the Solar Refining Company. Patents flew out of the Cleveland lab on refining sour oil in the last quarter of the year and during 1889. With Frasch's technology in hand, the value of the Ohio crude jumped in value from $0.14 per barrel to $1.00 per barrel. Considering that the Lima fields were producing 90,000 barrels per day, the Frasch technology increased annual revenues by more than $28 million per year! Without Frasch, Standard Oil may have never survived the gamble of the sour crude. Without Frasch, Imperial Oil did not survive.

What Frasch's partner Minhinnick thought of this conflict of interest, if he even knew, is not recorded. All of the experiments had been conducted at the Empire Oil refinery, and their partnership was but a year old when he sold out to Standard Oil. His ties with Empire Oil did not completely end until May 1888, almost two years after he had cut his deal with Standard Oil. There are many unanswered questions about Frasch working for Empire Oil and Standard Oil at the same time. Despite the fact that the title of the Frasch patent was "Refining

Canadian and Similar Petroleum Oils," the "Canadians" never even got rights to use the technology until after Standard Oil had taken over Imperial Oil in 1898. Even Empire Oil had to stop using the Frasch process. One wonders if he truly stopped being an employee of Standard Oil (through MMPC). Did Standard Oil set up the whole thing? Without doubt, the Empire Scandal contributed to the demise of Imperial Oil. Not only had they lost the technology to their rivals, when Standard Oil started to pressure Imperial in Eastern Canada, they, of course, were able to bring the Frasch technology with them. One wonders if Frasch had stayed loyal to his Canadian partners, if Imperial would have taken over Standard Oil! Knowing Frasch, he would have barked back that loyalty had nothing to do with it; it was all a matter of business.

Throughout his career, Frasch's actions are that of an independent consultant. He supplied what he was directly paid for, but anything beyond the defined scope was his to do with as he saw fit, at least in his mind. Whether his employers, clients and partners had formally agreed to such a relationship is unknown. They certainly were not expecting his ancillary endeavors to include their direct competitors.

Even Standard Oil, though, must have been concerned about their technology guru. At the same time, entrepreneurial ventures, outside the direct business of Standard Oil, were encouraged among the executive ranks of the oil company. Even so, Standard Oil had formally defined his "free time." This situation was perfect for the hyper-ambitious Frasch. Officially, he was an executive of theirs, and certainly most of his attention was given to his primary duties. He was also involved in various other enterprises, including ventures on soda ash (twice) and salt, and indirectly on linseed oil, enhanced oil recovery, gold mining and sulfur. Usually, these ventures were undertaken with other executives of Standard Oil, but without the direct participation of the company.

As the technology was being implemented at the Solar Refining Company, a new scientist joined Frasch's staff, William M. Burton (1865 - 1954), a recent graduate of John Hopkins University. Burton, himself a brilliant chemist, would discover petroleum cracking. He, too, would win a Perkin Medal (1922), and would become president of Standard Oil of Indiana. Burton was the most famous of Frasch's protégés.

Frasch's creative mind continued to work on petroleum. Between 1880 and 1900, thirty-four patents were filed by Standard Oil, half of which originated with Frasch. In 1895, another major invention was made by Frasch, the acidizing of an oil well to increase production, but this was not for Standard. Patents for the process were issued the following year (U.S. patent 556,651, 556,669). The acid reacted with the limestone rock, releasing carbon dioxide gas that opened up fissures in the rock. This process freed trapped oil pockets. Frasch had worked on this project with John W. Van Dyke (1849 - 1939). Frasch had known Van Dyke from the Solar Refinery years, where the latter had been the Superintendent of the revolutionary refinery for sour oil. On April 1, 1896, they assigned the patents to Van Dyke's company, the Oil Well Acid Treatment Company of Lima, Ohio. Van Dyke later led the Standard Oil spin-off company, the Atlantic Refining Company (later becoming part of ARCO and now part of BP Amoco).

In 1899, Standard Oil gave Frasch a new contract that gave him even more freedom than he had before. Now, he was only a consultant to the company for special projects. This gave him more time to spend running Union Sulfur. His last studies for the oil giant were on new types of oil. First, Frasch developed a process to refine the heavy aromatic oils from the Coaliga field in California. The patent for the process was filed on June 30, 1902 (U.S. Patent 968,760). Frasch's last patent for Standard Oil was filed on October 4, 1902 (U.S. patent 951,272). He had invented a new refining process for Texas oil. This "asphalt-based" oil was sour, but the sulfur was incorporated in a different manner than the "paraffin-based" oils that he had previously worked on in Ontario and Ohio. Coincidently, the discovery of oil in Beaumont, TX, had saved his sulfur venture that he had worked so diligently on for the past decade. He officially cut all ties with Standard Oil in 1905, after almost twenty years with the legendary petroleum giant.

4.2.3 HIS ALKALI VENTURES

In 1879, MMPC wanted Frasch to look at the manufacture of alkali. The petroleum industry, including Standard Oil, were major consumers, but there was no domestic supply (the Leblanc process never operated in North American and the first Solvay plant was not built until 1884 in Syracuse, NY). Frasch focused on improving the basic Solvay route to produce alkali. He may have first learned about the process at the Centennial Exhibition in Philadelphia in 1876, where Solvay displayed their state-of-the-art technology (a Sicilian sulfur display was also there). Construction began the following year, based on the technology developed by Frasch (U.S. patent 363,952), who was the design engineer for the plant. The American Chemical Company began operations at West Bay City, MI, on December 1, 1881. Frasch was the superintendent of the facility. Other executives of the firm were Edward Morgan (president), Joseph Meriam (vice-president), W.H. Morgan (secretary) and J.D. Ketchum (general manager). This plant was the first Solvay-soda-ash plant in North America, three years even before the Solvay Process Company (later acquired by Allied Chemical) built one. However, the American Chemical Company could not compete against the "real" Solvay plant once it was built, and the Frasch alkali plant was closed by MMPC in 1887. On August 17, 1888, MMCP sold the Frasch soda patent to their Solvay competition for $1,000. Frasch's soda exploits were not yet over. A new venture would arise after he came back from Ontario.

While in Ontario, three new soda patents had been filed by Frasch (U.S. patent 361,355; 361,622; 418,315). Returning to the U.S., he formed the Frasch Process Soda Company, with Frank and John Rockefeller as investors. His new Cleveland facility opened in 1892. Herman was the president of the company and he was joined by F. B. Squire. His second soda venture also eventually failed. Just as his alkali operations ceased in 1905, his most successful venture was starting to bring in significant revenue and profits.

4.2.4 HIS SALT VENTURES

The raw material for alkali was salt, and nearby deposits were one of the reasons for choosing the site for the American Chemical Company. Frasch filed his first salt patent on August 29, 1882 (U.S. patent 277,418), followed by an improvement patent filed on January 12, 1883 (U.S. patent 277,419). Rights to the patents were assigned by Frasch to MMPC. To supply the second alkali venture, the Frasch Process Soda Company, with salt, the group formed the United Salt Company in 1892, with Squire as president. One of Frasch's later salt patents (U.S. patent 1,006,196) was assigned to the United Salt Company. United Salt had two operations in Cleveland and one at Newburgh. The main Cleveland facility was badly damaged by a fire in August 1894. The financial performance of the company was not very good. Frasch had to take out personal loans from John Rockefeller to cover the losses. Frasch and his partners sold their interests in the company to the National Salt Company in September 1899. National Salt went into bankruptcy shortly afterwards and was reorganized as the Union Salt Company.

4.2.5 HIS SULFUR VENTURES: UNION SULFUR

The developments by Frasch in the petroleum industry were already enough to earn him, at least, a footnote in the history of the technical developments of industry. Frasch, though, switched from "Texas tea" to "Louisiana lemonade," and now his fame was assured as he created not just a new technology but a world-scale industry. While great inventions can lead to new global industries, seldom do inventors, themselves, create such industries directly.

Frasch's first great invention had involved removing sulfur. His second, even greater, discovery involved recovering the same element. He had removed sulfur from oil and recovered it from deep in the ground in Calcasieu, LA. The latter idea was brilliant but simple (an important trait of Frasch's inventions): melt the sulfur and then force the liquid to the surface with compressed air. With this "in-situ" method, a standard mine never had to be dug; only a well had to be drilled. Frasch had essentially converted the deep sulfur deposit into a giant subterranean *Calcaroni*, with no environmental discharges. While the concept was straightforward in principal, the practical application stretched Frasch's skills and determination (and his financial resources) to their limits.

While sulfur appears to be distant from his earlier research, it was closely related. According to Frasch [*J. Ind. Eng. Chem.*, **4** (2), p. 137 (1912)], his: "great deal of experience in drilling and mining petroleum and salt" were combined to develop a technology to remove sulfur from deep in the ground. Salt was especially relevant to this study. Salt deposits were "mined" by dissolving them in water and pumping the saline liquid to the surface. With sulfur, it was not so easy to dissolve (though Frasch did look into this), but it could be melted. His interests in sulfur may have been first aroused by the Dickert and Myers Company, whose head office was in Cleveland. The company was owned by Ferdinand Dickert (~1841 - 1907) and Daniel Myers, who was the president of the company. On March 5, 1881, they had sought Rockefeller's help with their struggling sulfur mining operations in Cove Cree, UT. Dickert later patented a process to refine sulfur with superheated water (*U.S. patent 298,734, 301,222, 1884*). Frasch had also heard about the ventures down in southern Louisiana.

With Calcasieu in mind, Frasch applied his creative genius to come up with a solution to the alluring, but distant, treasure. Without ever visiting the site, or drilling one well for sulfur, he filed his sulfur patent on October 23, 1890 (US patent 461429; see Figure 4.2, "H. Frasch Mining Sulfur" [title page and 1st page of patent]; 461430; see Figure 4.3). Never an arm-chair inventor, Frasch still had to convert this theoretical exercise into a proven practical process. To legitimize his paper technology, on November 12, 1890, he enlisted two of his trusted colleagues, Frank Rockefeller and Feargus B. Squire (1850 - 1932), a co-founder of Standard Oil; both had been associated with the Solar Refining Company and the Frasch Process Soda Company. In return for their financing, they shared ownership of the sulfur patents, 30% to each.

The new syndicate had a bit of a conundrum on their hands. The sulfur at Calcasieu was under the control of the American Sulfur Company (ASC). Frasch did not particularly want to share his discovery with anyone outside of his group. His strategy was to work around them. Frasch presumed that the sulfur deposit extended outside their land lease. He was wrong. In 1891 and 1892, Frasch and his partners had commissioned four exploratory wells and had found no sulfur. ASC must have wondered what the naive newcomers were doing; they soon found out. Frasch went to New York and met with Edward Cooper of ASC in early 1892.

Cooper was skeptical, but had few other options for the property. ASC hired the mining engineer Rossiter W. Raymond (1840 - 1908), who would later be president of the Institute of Mining Engineers, to evaluate the technology. A positive response to the "radical" technology led to an agreement for a partnership between ASC and the Frasch syndicate, but only if Frasch could prove his untested technology. A later article (*New York Times*, Dec. 18, p. 4, 1910) stated that Twombly had been the first to become interested in the project and that he had

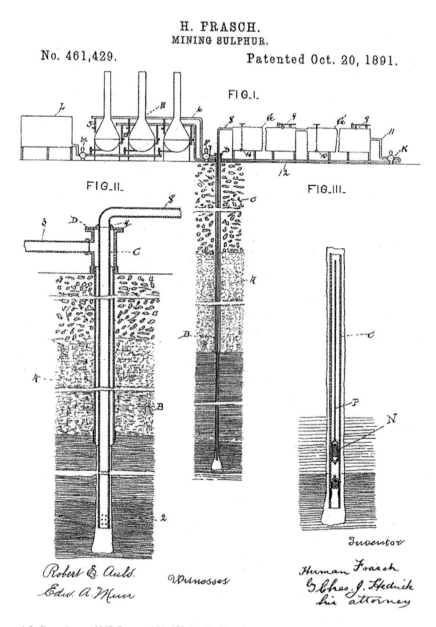

Figure 4.2. Drawings of US Patent 461,429 by H. Frasch.

introduced Frasch to Abram Hewitt. An earlier report (*New York Times*, Oct. 12, p. 1, 1895) stated that the Vanderbilts, whom Twombly was related to by marriage, were part of the original Frasch syndicate, so there may be some truth in the Twombly connection.

Frasch's sulfur exploits took place during his "vacation" time from Standard Oil. His first of many sulfur holidays took place in early 1894 for the drilling of the first Frasch test well (#14 on the site) on the ASC property. In September 1894, Frasch sent his colleague, Jacob C. Hoffman, down to Louisiana on his vacation (where it became his full-time job) to set up the drilling operation. For the trials of the technology, Hoffman was Frasch's front man and did most of the dirty work. Hoffman was ably aided by the blacksmith Jacques Toniette (~1859 - 1924). Toniette and his wife had been born in France, but had begun raising their family in Marmora, ON. In 1890, he had been brought down from Canada by ASC to handle the pipe fitting and general metalworking duties.

In early December, the ten-inch pipe reached the elusive bottom of the sulfur layer at Calcasieu. Until this time, Frasch had remained in Cleveland. He did not arrive at the mine until mid-December. There were delays and Frasch returned to Cleveland for Christmas, but came back a few days later. The first "Frasch" trial took place just before New Year (the exact day is not recorded). After adding superheated water to the well (#14) for 24 hours, sulfur was pumped out for four hours, **the most famous technological milestone in the history of the sulfur industry**. Frasch spoke of the excitement of the first flow of sulfur [*J. Ind. Eng. Chem.*, **4** (2), p. 137, 138 (1912)]:

> ...a beautiful stream of the golden fluid shot into the barrels...I enjoyed all by myself this demonstration of success. I mounted the sulfur pile and seated myself on the very top...Many days and many years intervened before financial success was assured, but the first step towards the ultimate goal had been achieved.

Sulfur was, at last, recovered from Calcasieu, a quarter century after the first attempts had been made by CSM. What others had spent decades trying to do, Frasch did in a matter of months.

While a great technical milestone, a long and arduous path still lay ahead. He had taken but the first step in the long and winding staircase of commercial ventures. Frasch returned to Cleveland to his executive position at Standard Oil. After he left, technical problems caused the liquid sulfur to solidify in the pipe of the well. The greatest fear of a Frasch mine had taken place just after its first trial. The drilling to unplug the pipe took months. More than five years had passed since Frasch had filed his patents. More than another five years would pass before the company and technology would be commercially successful. Up to this time, the

UNITED STATES PATENT OFFICE.

HERMAN FRASCH, OF CLEVELAND, OHIO.

APPARATUS FOR MINING SULPHUR.

SPECIFICATION forming part of Letters Patent No. 461,430, dated October 20, 1891.

Application filed December 26, 1890. Serial No. 375,799. (No model.)

To all whom it may concern:

Be it known that I, HERMAN FRASCH, a citizen of the United States, residing at Cleveland, in the county of Cuyahoga and State of Ohio, have invented certain new and useful Improvements in Apparatus for Mining Sulphur, of which the following specification is a full, clear, and exact description.

This invention relates to the removal of sulphur from deposits or mines in the earth which consist of or contain free sulphur, and is particularly useful in the removal of the sulphur from deposits which are overlaid with beds of quicksand, and which therefore cannot be mined in the usual way by sinking a shaft; but each of the improvements constituting said invention is included for all the uses to which it may be adapted. By means of the present invention, moreover, a refining of the sulphur is effected in the mining operation.

The present invention consists in apparatus whereby the removal of the sulphur from the mine or underground deposit in a liquefied state is effected or facilitated.

In accordance with the said invention a well is sunk into or through the underground sulphur deposit or mine, and a pump or other known or suitable means of forcing circulating or elevating liquids is employed to remove the liquefied sulphur or the liquefied sulphur and vehicle by which it is liquefied. Heaters are also used for raising the temperature of the liquefying-vehicle; also vessels for recovering the sulphur after it has been removed in a liquid state, together with other appliances, as hereinafter set forth.

In the accompanying drawings, which form part of this specification, Figure I is a diagram of a plant for mining sulphur in accordance with the invention. Fig. II is an enlarged diagram of the well. Fig. III is a view illustrating a modified arrangement of part of the apparatus. Fig. IV is a view illustrating a further modification. Figs. V and VII are each a diagram of a plant for mining the sulphur by means of a solvent vehicle, the form of the apparatus exhibited in Fig. I being especially adapted for use with a fusing-vehicle; and Fig. VI is a view of a modified arrangement of part of the apparatus of Fig. V.

Referring to Figs. I and II, a well A is drilled, as usual, in making salt and oil wells, a casing B (say ten inches in diameter) being brought to the rock above the sulphur, so as to shut off water and quicksand, and a smaller hole (say eight inches in diameter) being continued into or to the bottom of the sulphur deposit. Tubing C (say five inches in diameter) is introduced through the sulphur and sulphur-bearing rock nearly to the bottom of the well. The lower part of this tube may be provided with perforations at the side, or may otherwise be provided with a strainer, or it may be left more or less open. There are thus two passages opening at the bottom into the underground deposit. At the surface there is a casing-head D, having a lateral tube 3. The tubing C extends through the casing-head D. At E are boilers or hot-water heaters, which may be used to heat water under pressure to a suitable temperature to melt the sulphur when forced into the mine at the bottom of the well. A temperature of from 270° to 280° Fahrenheit, which requires a pressure of about thirty pounds to thirty-five pounds per square inch, will suffice with an appropriate flow. The water-spaces of the heaters are connected by branches 5 and a main 6 with the inlet of a force-pump F, whose outlet is connected by a pipe 7 with the tube 3 of the casing-head D. The pump has sufficient capacity to force a stream of water through the downflow-passage, consisting of the casing B and the hole in continuation thereof, and through the upflow-passage of the tubing C with less loss in temperature than will bring the water down to the melting-point of sulphur—say with a loss of about 35° Fahrenheit, supposing the water to enter the casing at a temperature of 280° Fahrenheit. The flow of the water and its temperature should be such that in the sulphur or sulphur-bearing rock and in passing up through the tubing C it remains at a temperature at which the sulphur is liquid. The pipe 8 forms a continuation of the tubing C to the sulphur-recovering vessels in the form of settlers, of which a series G G' is shown, each settler or settling-tank being closed and provided with a safety-valve 9, loaded to, say, sixty-five pounds per square inch, and also with a draw-off 10 at the bottom of the sulphur. From the last settler a pipe 11 con-

Figure 4.3. The first page of US Patent 461,430 by H. Frasch.

sulfur project was but a "hobby," to use Frasch's own words, as his main (and paying) job was still with Standard Oil.

They were not ready for the second test until September of 1895. Frasch was there again on vacation. The second test went well, producing 500 tonnes of sulfur. These results were enough to convince ASC to merge with Frasch and his partners. With each group owning 50% of the company, the Union Sulfur Company was formed on January 23, 1896. The company's first president was Alexander Tiers. Frasch was superintendent; he did not become president until after he resigned from Standard Oil in 1899; at which time, Tiers became Treasurer. On May 18, 1896, the Frasch patents were officially assigned to the new company.

To increase production, a new pumping station and a larger boiler house had to be installed. During this period, Frasch came up with another brilliant, but simple, idea of using compressed air to force up the liquid sulfur, instead of mechanically pumping it. Failure of the pumping system had ended the first two trials. The new idea proved extremely successful in the third trial, the first for the newly formed company: April 19th to May 26th - 2,000 tonnes. Just when results were looking good, they got stuck in a rut. New wells were drilled. Well #15 only produced 150 tonnes in two trials, and well #16 was destroyed before producing any sulfur. Well #17 was more successful, from February 8 to March 29, 1897 producing 1,115 tonnes. The same well yielded 1,343 tonnes in another trial in early 1898. In this year, well #18 was drilled, but it was also a failure, producing only 41 tonnes. Frasch could just not get the wells to produce consistently. Union Sulfur seemed to be so close to success; at the same time, so close to bankruptcy.

In 1896, total sales had been $23,293, which barely covered the cost of fuel for the hot water boilers. Cash reserves had been depleted by the end of the year. By September 1897, there had been only another $16,000 in sales, while the company owed $320,000! Operations were suspended in the fall of 1898, after producing 1,484 tonnes of sulfur, and no further activity took place at Sulfur Mine in the remainder of 1898 and 1899.

Even the determined Frasch had doubts. Another one of his ventures looked like it would fail as well. In a desperate move, he decided to change his strategy. In late summer of 1898, Frasch took his vacation in Sicily to try his process over there. Again, Frasch returned to Cleveland and let poor Hoffman do the dirty work. The Sicilian mission failed miserably. A well was drilled 1,000 feet into the ground and no deep deposits of sulfur were found. Another well also yielded no sulfur. The dejected and abused Hoffman had enough and resigned in April 1899, after the failure of the year-long Sicilian sulfur folly.

Frasch still passionately believed in his technology. Unlike his other ventures, Union Sulfur used a technology of his that was unique. The potential upside was staggering, if only… Determined as ever, Frasch himself proclaimed [*J. Ind. Eng. Chem.*, **4** (2), p. 137, 138 (1912)]:

> This severe criticism, while not agreeable, did not carry very much weight with me. I felt that I had given the subject more thought that my critics, and I went about my work as best I could, thoroughly convinced that he who laughs last, laughs best.

The year 1899 was a pivotal point in the history of Union Sulfur. Frasch had to make a decision on where he was going to focus his energies. He chose sulfur. He stopped his endless search for new ventures. His own company now provided all the excitement that he needed in his business life, and the problems at Calcasieu required his full-time attention. Unlike many entrepreneurs who pioneer new technology, Frasch had an easy out. He was still an executive at Standard Oil. At the time, his career was with petroleum; sulfur had been but an interesting diversion. The average person would have returned to the corporate safety net in Cleveland and would have found better places to spend their vacation. Frasch did just the opposite. In 1899, he negotiated a new contract with Standard Oil that gave him more time to focus on his sulfur project. Frasch then boldly went back to his partners looking for more money. He believed that the mine could be successful if three or four wells were operating at the same time. To do so, he needed another $600,000, a substantial sum as the end of the century approached. How he convinced them under the daunting circumstances is unclear. Five of his backers had already withdrawn. Worst of all, the problem was not only the production rate, but that the cost of energy was too high. He had no idea how to slash these costs. Energy was the Achilles' heel of the Frasch process (and was ultimately responsible for its demise).

Now, ten years from the patent filings, activity slowly resumed at Calcasieu. Only 36 tonnes of sulfur were produced in 1900. Union Sulfur was not in good shape at the start of the new century. The company owed creditors $138,653. Since the inception of Union Sulfur, consolidated sulfur sales only added up to $67,327.

Adding to their woes, the largest sulfur market, the manufacture of sulfuric acid, had disappeared in the U.S. When Frasch had filed his patent over 80% of the rapidly growing American sulfuric acid industry used elemental sulfur as its raw material. By 1901, 85% of the industry had switched to pyrites. The only major acid producers using elemental sulfur were Kalbfleisch (later purchased by Cyanamid in 1929) and Grasselli Chemical Company of Cleveland. Most of the sales of Union Sulfur had been through Petit & Parsons in New York City for resale. Other shipments were made to a sulfite pulp mill in Wisconsin and fertilizer manufacturer in Meridian, MS.

The perfect storm of the sulfur world had struck Frasch and Union Sulfur. Another group of investors appeared to be destined to succumb to the golden lure of Calcasieu. Only a miracle could save Union Sulfur, and a miracle they got! On January 10, 1901, oil was struck by Captain Anthony Lucas (1855 - 1921; his real name was Antonio Francesco Luchic) at Spindletop, near Beaumont, TX, which was only sixty miles from Sulfur Mine. This gusher was one of the greatest oil finds in the history of the petroleum industry, and it was in the backyard of Sulfur Mine. Energy costs for hot water dropped by more than a magnitude overnight.

With the energy woes behind them, Frasch now had a chance, but he still had to get production up. Union Sulfur was still far from being a successful company. In 1901, Union Sulfur produced 3,000 tonnes, better but still far from what was needed. The production was enough for Union Sulfur to be the largest producer of sulfur in the U.S. (which did not say too much). Sulfur sales were only a paltry 900 tonnes, valued at $12,000, in this year, and inventories stood at 2,000 tonnes. The next year was not much better, as the drilling of a new well (#22) failed and total production only increased to 5,000 tonnes. Union Sulfur was still teetering on bankruptcy.

Others attempted to "undermine" the pending success of Union Sulfur. In 1902 and 1903, several ventures tried and failed in the area using non-Frasch methods, including the Gladys Oil and Sulfur Company, the Louisiana Mining and Oil Company, the Vinton Oil and Sulfur Company, the Union Oil and Sulfur Company, and the Dirigo Oil and Sulfur Company.

The year 1903 was the break-out year for Union Sulfur. The summer of 1903 saw four new wells, and production had risen to almost 25,000 tonnes. Union Sulfur was set to explode on the world sulfur scene. The first full year of success followed. In 1904, they were the second largest sulfur producer in the world, with 80,000 tonnes of production. This year is **the most famous business milestone in the history of the sulfur industry**. The Frasch technology had finally arrived. They had come a long way in the six years since the "experts" had written them off. Frasch no longer needed his security blanket back in Cleveland. He cut his ties with Standard Oil to focus on being president of his company. Union Sulfur was set to take on the world.

Those that had stayed with the company over the years became very wealthy from their investments. The major shareholders (90%) of Union Sulfur were:

- Herman Frasch (he had also purchased the shares of Squire);
- Louis H. Severance (~1848 - 1915) – Frasch's partner, who was treasurer of Standard Oil and Union Sulfur; he had purchased the shares of Frank Rockefeller;
- Hamilton McKown Twombly – a partner in ASC (2nd largest shareholder after Frasch);
- Abram Stevens Hewitt – a partner in ASC;
- Edward Cooper – a partner in ASC.

In 1905, Union Sulfur produced more than 200,000 tonnes of sulfur from Calcasieu. Over six hundred people were employed at the mine. The Louisiana sulfur first arrived in the New York market in July 1905 and had displaced Sicilian imports by the fall. By 1907, 30 sulfur wells were active. The sulfur world was forever changed. Not only was there a new producer, this one mine was producing as much as a hundred mines in Sicily at a fraction of the cost.

Now that Union Sulfur was on firm financial footing, it was time to put the Sicilians in their diminutive place. The ASSC was at a serious cost-disadvantage compared to Union Sulfur, and Frasch knew it. Frasch wasted no time in going to London to tell the ASSC that they had better back off from the U.S. The ASSC directors were not impressed with the threats of the German-American. Since they would not listen to reason, Frasch would have to show them the hard way. Upon returning to the U.S., he "bombed" the price in the U.S. If the Sicilians still had any doubts, they were laid to rest when, in November 1904, a 3,000 tonne shipment landed in their backyard at Marseilles (the first U.S. exports to Europe). Union Sulfur first chartered vessels, and, in December 1909, launched its own steamer, the *Herman Frasch* (5,000 tonne capacity; sunk in a collision on October 4, 1918, while being used by the U.S. Navy as a troop carrier). The young grandson of Herman, Herman Frasch Whiton had christened the ship. The lost vessel was replaced by a second *Herman Frasch*, launched on June 5, 1920, and was christened by Frieda Whiton. The fashion of the day was to name such company vessels after the executives of the firm, and, at times, their family. In 1913, the company launched a second vessel, the *Frieda* in February (torpedoed in 1943). Later, the *Severance* (decommissioned in 1919), the *Hewitt* (disappeared off the coast of New Jersey on February 1, 1921), and the *Henry D. Whiton* in 1921, were launched.

The shock of the Marseilles shipment changed the minds of the directors of ASSC. Union Sulfur withdrew from Europe after reaching an agreement with ASSC, whereby Sicilian exports to the U.S. were restricted to 75,000 tonnes per year. However, ASSC dissolved shortly afterwards. Frasch now had to make his point all over again. Frasch placed the hard-nosed Herman Hoeckel as his European director. Hoeckel led the initial negotiations with the new Sicilian cartel under COISS, but the negotiations quickly bogged down. Frasch stepped in. In 1908, a new agreement was hashed out to divide the sulfur world. The agreement collapsed in 1912 after new U.S. laws made the legality of the arrangement questionable. By this time, Frasch was too ill to be involved in the day-to-day running of Union Sulfur or future negotiations. One of his last acts as head of Union Sulfur was in January 1913, when he officially informed COISS that he was canceling his trade agreement with them. By this time, Frasch production surpassed that of all of Italy to become the largest producer of sulfur in the world. The U.S. has retained this position ever since!

An integral aspect of the Frasch strategy for dealing with COISS had been a ghost plant; the facility was built but never operated. Union Sulfur constructed a massive sulfur refinery in Marseilles: Raffineries de Soufre Internationales (RSI), allegedly to produce highly-pure, "flowers of sulfur," which was used as a fungicide on grapes. COISS had convinced French customs to give Frasch's imports a hard time. The plant, with a capacity of 100,000 tonnes per year, had been constructed as a compromise with local authorities to allow American sulfur into Marseilles without tariffs. The new plant in Marseilles, its mission accomplished, was never operated! The empty plant later came back to haunt Frasch and especially Hoeckel. Frasch wanted to be convincing, and RSI was built like a fortress. During World War I, the French authorities became suspicious about the massive plant built by a German for a company founded by a German and not producing anything in the heart of Marseilles. French authorities accused Hoeckel of being a spy. The rattled Hoeckel fled France and was never heard from again.

The Marseilles Fortress had not been the first intrigue that Frasch had been suspected of. The ailing Frasch was claimed to be indirectly responsible for the Italo-Turkish War over Libya which lasted from September 1911 to October 1912. In March 1911, the Italians had accused an American archeological expedition of clandestinely investigating the Tripoli sulfur situation. The Italians resented that the Americans had been given permission to study the ruins of Cyrene, when their request had been turned down. The press played up the reports, and the government had already been making plans to annex Libya. A naval expedition had been under consideration by the Italian government at the time. Making matters worse, Union Sulfur then gained control of the sulfur deposits in Tripoli. Frasch had obtained a concession granted by the Sultan of the Ottoman Empire. The Italian government saw the new nearby sulfur deposit as a serious threat to their Sicilian business. They were looking for an excuse to declare war against the Ottomans, and this was as good as any. They decided to take the sulfur deposits and all of Libya with it by force. While the story sounds like a tabloid report, the source, published in the *New York Times* (May 3, 1914), was claimed to have been Frasch himself. This conflict should have been called the "Sulfur War of 1912."

Frasch's aging technical skills were required by Union Sulfur one last time. Sulfur Mine had suddenly stopped producing. Frasch, who was in Europe at the time recuperating, returned and resolved the problem in one of his last visits in November 1911. After all these years, the water build up was cooling the fresh hot water being added. Bleed pumps were installed to remove the standing water. By this time, one Frasch well could produce 400 tonnes per day. The technology had come far in the past decade. After Frasch came to the rescue, Union Sulfur produced more than 775,000 tonnes of sulfur the following year. Production in the early years of the company is summarized in Table 4.1 (also see Figure 4.4).

While Frasch's technical developments and general business savvy were widely acclaimed, his confidence, at least in one case, got the best of him. Earlier, Union Sulfur, while still under the protection of the Frasch patents, had been offered the rights to the sulfur domes at Bryanmound and Big Hill. Frasch turned them down, stating (Haynes, Volume II, p. 207):

> Gentlemen. I am not interested. Nobody else can pump sulfur as we do, and in Louisiana we have all the sulfur we need for years and years. Someday, if I should ever want your property, I will buy it at grassland prices.

Frasch had reason to be overly confident at the time. However, his words would later come back to haunt Union Sulfur. The original Frasch patent was expiring in 1908. To extend his protection, a series of improvement patents had been filed earlier on October 30, 1903 (U.S. Patent 870, 620; 988,994; 988,995). Two more followed on February 6, 1905 (U.S. Patent 928,036; 1,008,319). Although now ailing, his last patent was filed on May 3, 1912 (U.S. Patent 1,152,499). Despite these later filings, six months after the last patent, another company had started using his technology to mine sulfur. Swenson, the president of Freeport Sulfur, had offered to pay a royalty to Union Sulfur, but the offer was refused. In 1915, Union Sulfur sued Freeport Sulfur for violating the improvement patents. The company had waited until the last of these patents had been issued (U.S. Patent 1,152,499; issued September 7th). Freeport Sulfur maintained that the pioneering patent had expired and the improvement patents were not inventive. A lower court decision had gone in favor of Union Sulfur. However, in October 1918, the U.S. Circuit Court of Appeals ruled in favor of Freeport Sulfur. Later, Union Sulfur also took action against Texas Gulf Sulfur. Their case was settled in 1921. These two companies would become the two largest sulfur companies in the world by the late 1920's. By this time, Union Sulfur was no longer in the sulfur business because its sulfur reserves were depleted. One wonders what the courts would have decided if Frasch himself had been there to defend his work.

By October 1917, Union Sulfur had produced 4,682,000 tonnes of sulfur. The company was rapidly going through the sulfur reserves at Calcasieu. Union Sulfur now regretted not taking advantage of the earlier offers to invest in Big Hill and Bryanmound; this had been one of Frasch's few errors in judgment. Not until 1923 did they begin drilling new wells. By then it was too late. Over fourteen new domes were drilled without success. On December 23, 1924, two days before the 73rd anniversary of the birth of their founder, Sulfur Mine closed, after producing almost ten million tonnes in its history, and on May 7, 1926, six days after the twelfth anniversary of the death of Frasch, the company made its last shipment (from inventory) overseas. Williams Haynes described the situation (**The Stone That Burns**, p. 237):

> The ghost of the defunct Frasch mine in Calcasieu Parish haunts the management of every sulfur-mining business.

In 1928, Union Sulfur attempted to re-enter the sulfur business. On November 14th, wells were drilled at Boling Dome, but Texas Gulf Sulfur out bid its competitors to the sulfur rights to the lucrative site. The sulfur years were not yet completely over for Union Sulfur. For decades the company retained the rights to the abandoned Sulfur Mine. Like a yellow phoenix, Sulfur Mine was reopened in 1967, the centennial of the discovery of sulfur at the site, by Union Sulfur now operating under the name Union Texas Petroleum (owned by Allied Chemical). Sulfur Mine closed permanently three years later.

Union Texas Petroleum was the last of the successor companies to Union Sulfur. Union Sulfur did not fold when its sulfur reserves were depleted. When the mine first closed, Union Sulfur prospered becoming an oil and gas exploration and production company. A major source of revenue to the company was the oil reserves found at Sulfur Mine. For a long time they kept the name Union Sulfur even though no longer in the sulfur business. The company name later changed several times–Union Sulfur and Oil in 1950; Union Oil and Gas in 1955; Union Texas Natural Gas in 1960–before becoming part of Allied Chemical in 1962. Allied Chemical spun off the division in 1992, through an $840 million IPO. Union Texas Petroleum had sales of one billion dollars when taken over by ARCO (now part of BP-Amoco) in May 1998 for $3.3 billion. This is what had become of the dream of Herman Frasch.

Table 4.1. Frasch company summary – Union Sulfur production from 1894 to 1915 (1914 not available) in tonnes

1894	1895	1896	1897	1898	1899	1900
5	500	2,101	1,145	1,384	0	36

1901	1902	1903	1904	1905	1906	1907
3,078	4,983	23,715	79,187	218,950	287,590	188,878

1908	1909	1910	1911	1912	1913	1915
367,896	273,363	247,060	204,220	786,605	491,080	475,128

Source: Williams Haynes, **Sulfur the Stone that Burns**;
Sutton, W.R.: **Herman Frasch**, Dissertation, *Louisiana State University*, May 1984.

4.3 THE OTHER SULFUR COMPANIES

A great sulfur industry had been created by Herman Frasch, with Union Sulfur being the first member of this elite group. However, the time of this company, at least in the sulfur business, was relatively limited.

The salt dome formation at Calcasieu was not unique. These geological formations were scattered across the Gulf coast of Texas and Louisiana. Among the salt domes, about 15% of them were also found to contain commercial quantities of sulfur, usually on the cap-rock covering the salt deposit. Most of the sulfur finds were incidental discoveries when drilling for oil, which had also been found in these formations. While the yellow, instead of black, discovery was a disappointment to the oil drillers, they provided opportunities for those interested in sulfur. A pattern developed where many of the future Frasch domes were discovered by oil exploration companies, which leased or sold the rights to the highest bidding Frasch company. The oil companies often did very well out of these "failed" wells.

After the basic Frasch pioneering patent expired in 1908, new "Frasch" companies entered the marketplace. The first of the newcomers was the Freeport Sulfur Company, whose mine at Bryantmound, TX, opened on November 12, 1912. In 1917, the U.S., through the production of Union Sulfur and Freeport Sulfur, became the first country to produce over one million tonnes of sulfur in a single year. Production was double the record rate in Italy, which had been set in 1905. Italy remained the second largest producer followed by Japan.

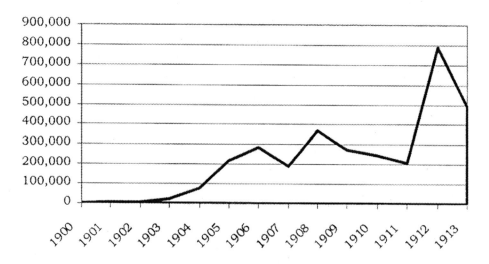

Figure 4.4. Frasch company summary — Union Sulfur. Production from 1900 to 1913 in tonnes.

On March 19, 1919, the other major sulfur company, Texas Gulf Sulfur, began operations at the Big Hill Frasch mine. By the mid-1920's, the sulfur reserves at Calcasieu were depleted and Union Sulfur withdrew from the sulfur business. Two "junior" sulfur companies, Duval Texas Sulfur Company in 1928 and Jefferson Lake Sulfur Company in 1932, entered the Frasch industry. Duval and Jefferson Lake usually mined smaller domes, often ones that Freeport and Texas Gulf had passed over. Freeport and Texas Gulf remained the dominant sulfur producers. Their sulfur production numbers from the beginning through the glory years of the Frasch industry are summarized below (see Table 4.2; and Figure 4.5).

Table 4.2. Frasch company summary – US sulfur production in 1000 tonnes

Company	1905	1915	1935	1940	1945	1950	1960	1965
Union Sulfur	219	479	0	0	0	0	0	0
Texas Gulf Sulfur	0	0	923	1,428	2,203	3,154	2,335	2,600
Freeport Sulfur	0	42	612	860	1,179	1,559	2,200	3,500
Duval Texas Sulfur	0	0	64	217	181	201	139	119
Jefferson Lake Sulfur	0	0	24	220	190	278	363	312
U.S. Total (*)	**219**	**521**	**1,633**	**2,732**	**3,753**	**5,193**	**5,037**	**6,116**

(*) Total is from the **U.S.G.S. Minerals Yearbook**, and not necessarily a summation of individual company production.

4.3.1 FREEPORT SULFUR

Sulfur had been found at Bryanmound in the summer of 1906, just after Union Sulfur had started making an international name for itself. This site was of historic significance, for Stephen Austin (1793 - 1836) had established his Texan colony here in 1832. The property was sold to his nephew William Joel Bryan (1815 - 1903) in 1860.

In 1908 (the year of the expiration of the pioneering Frasch patent), the Gulf Development Company was formed to exploit the sulfur dome at Bryanmound. The founders were Edward Simms, George Hamman (1874 - 1953) and Henry T. Staiti. The following year, J.P. Morgan (1837 - 1913) had been approached to invest in the under-funded project; he contacted Bernard Mannes Baruch (1870 - 1965) to take a look at it. The Morgan plan was that Baruch would carry out the evaluation and get 40% of the profits of the company while Morgan would provide the financing and get 60%. Baruch contacted the noted mining engineer, Seeley Mudd (1861 - 1924), to evaluate the project. Mudd felt that the project only had a 50% chance of success; that was not high enough for Morgan and he backed

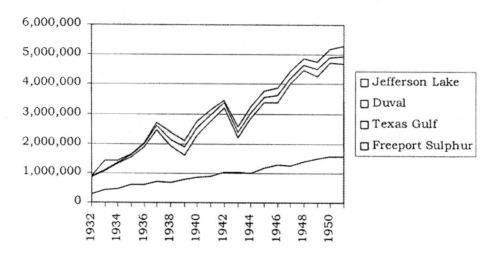

Figure 4.5. Frasch company summary — U.S. production from 1932 to 1951 in tonnes.

away from it. Union Sulfur was also offered the rights to the sulfur, but turned it down. The Simms-Hamman-Staiti syndicate had more luck with Eric Pierson Swenson (~1855 - 1945), the director of the First National City Bank of New York. His father, Swante Magnus Swenson (1816 - 1896) had immigrated to Texas about the same time as William Bryan; both of them had been a close associate of Sam Houston (1793 - 1863). Joining Swenson in the company were Sidell Tilghman (~1849 - 1927); Eric Swenson had married his sister, Maud Tilghman (~1861 - ?), in 1889; Frank A. Vanderlip (1864 - 1937), also from First National, William T. Andrews, Charles A. Jones (1861 - 1934), and Frank Hastings.

A new company, the Freeport Sulfur Company, was formed on July 12, 1912, with Eric Swenson as its first president. The company had been shrewd, hiring Benjamin Andrews, who had worked for Union Sulfur and knew the Frasch process well, as its operations general manager (he left in October 1913). On November 12th, the fresh Frasch mine, the second in the world and the first in Texas, began producing elemental sulfur. Nearby the town of Freeport was built to service the newly-opened Frasch mine. From a geological point of view, Bryanmound was a poor site for the Frasch process, requiring more hot water than Sulfur Mine. Production was only 10,000 tonnes in 1913, rising to 260,000 tonnes three years later. Total production of Bryanmound in its first five years was 884,000 tonnes. Closing in 1935, this Frasch mine had produced more than five million tonnes of sulfur.

The timing of the start-up of Freeport was fortunate for them and Union Sulfur. By then, demand was soaring and a sulfur shortage developed because of

World War I. During the later war years, Bryanmound peaked. Union Sulfur now had a real competitor.

Freeport Sulfur underwent a corporate coup that stands out in the chronicles of the business world. While effectively running the company for almost two decades, Swenson had alienated many of the shareholders, especially the Richmond banking firm, John L. Williams and Sons, by his secretive and controlling ways. This latter company, under the sons of the founder, John Skelton Williams (1865 - 1926) and Langbourne Meade Williams (1872 - 1931), had been an investor from the beginning and had a long-running feud with management. John Williams had begun the struggle, which was taken over by his brother upon his death in 1926. The issue had escalated to open warfare by the late 1920's. On October 16, 1928, Langbourne Williams Sr. wrote to the board that Swenson had a "lack of frankness and fair dealing." Swenson was also accused of deliberately wanting to see the stock value fall since he had recently sold some stock and wanted to buy back in at a lower price. The disgruntled shareholders were especially concerned that the sulfur reserves at Bryanmound were understated, which artificially depressed the share price. Requests to Swenson about the sulfur reserves and other critical information about Freeport Sulfur had been met with silence. John L. Williams & Sons, under the young Langbourne Meade Williams Jr. (1903 - 1994) of the family firm, led the proxy battle to oust Swenson. However, the Depression foiled his plans. Swenson fought back. In February 1930, he sued John L. Williams & Sons for $1 million for libel. Williams sought assistance from his one time boss, John Tyler Claiborne Jr. Claiborne convinced the even younger and very rich, John (Jock) Hay Whitney (1904 - 1982) to join them in the revolt. Whitney invested $500,000, becoming the company's largest shareholder. At the AGM on April 10, 1930, Swenson was removed as president.

The young executives, who had overthrown the established management with Machiavellian efficiency, were 32 years old (Williams) and 28 years old (Whitney). Two corporate babies, Williams and Whitney, now controlled the company. *Time Magazine* reported (April 2, 1934) that Freeport Sulfur had the youngest management of any U.S. company. They delayed taking direct power, because of their age. They brought in the Baltimore banker Eugene Levering Norton (1881 - 1960), as interim president and Odie Richard Seagraves (1886 - 1970), ex-president of Union Gas (owner of Duval Texas Sulfur), as chairman, while Williams became VP. In 1933, Williams moved into the position of president and Norton became chairman, but only for one year. In late March 1934, Whitney replaced Norton as chairman of Freeport Sulfur. Besides a stint overseas during World War II, Whitney remained chairman until 1958, when he became ambassador to Britain. Whitney was also noted as the founder of Pioneer Pictures, which introduced Technicolor to the movie world. Among conspiracy theorists, the two baby executives were later connected with CIA political intrigue in Cuba, Indonesia, and, even the assassination of President Kennedy.

After the closure of the Union Sulfur mine, all Frasch production had been in Texas. This situation changed in the early 1930's with the opening up of Grand Ecaille by Freeport Sulfur. The sulfur dome had been discovered by Shell, Humble (now part of Exxon) and Gulf during oil exploration. Freeport Sulfur purchased the rights and production began on December 8, 1933. By the end of the month, almost 18,000 tonnes of sulfur had already been produced. The new dome presented new challenges. Built in marsh lands, the rigs were placed on barges and canals dug to move them into position. The sulfur was contaminated by traces of petroleum. A novel distillation process was invented to purify the sulfur.

During the Depression, Freeport Sulfur never lost money (see Table 4.3a). Their only difficult times were the early 1920's, after the start-up of Texas Gulf Sulfur. Losses were posted in 1921, 1922, and 1924. In 1928, shares were selling for $105. In the 25-year period 1919 to 1943, the average profit-after-tax return was 14% on capital employed, and the accumulated sulfur production by Freeport Sulfur was 15 million tonnes. During the 1940's, the company was producing more than one million tonnes per year. Freeport Sulfur also opened a sulfur recovery plant from refinery gases in New Jersey in 1950, and during the sulfur crisis of 1951, they purchased pyrites deposits in Quebec from Arnora Gold Mines. By the early 1960's, Freeport Sulfur overtook Texas Gulf Sulfur as the largest sulfur producer in the world. During the middle part of this decade, the company was producing more than four million tonnes of sulfur per year.

Each new Frasch mine was generally more costly than the last, because of greater technical and engineering challenges. The exploration spread from marsh lands to tidal areas to off-shore into the Gulf itself. A milestone for the Frasch industry was the first off-shore dome to be mined. In 1960, Freeport Sulfur opened the Grand Isle Frasch mine (closed 1991), off the coast of Louisiana. The liquid sulfur was transported through a seven-mile underwater pipeline to shore installations. The pipeline consisted of a hot water jacket surrounding the sulfur pipeline, with the outer diameters of the pipeline being: 14 inches for the protective casing; 7 5/8 inches for the steam line; 6 inches for the sulfur line. The steam requirements were 75 MBTU/day.

Under the direction of Williams and Whitney, the sulfur business prospered and Freeport Sulfur started to diversify its product portfolio becoming a broader-based resource company. Freeport had entered the manganese business in 1931, with operations in Cuba (Cuban-American Manganese Corporation). At one point, they were the largest manganese producer in the world. This business was dissolved in 1946, after depletion of the ore. Other Cuban investments were the Nicaro Nickel Company in 1943 and the Moa Bay Mining Company, which were later confiscated by the Castro government. Nickel quickly became one of the largest businesses of the company. They moved into oil and gas. Freeport also acquired 47% ownership of the National Potash Company; and Southern Clays Inc. (becoming Freeport Kaolin) in the early 1960's.

Williams remained president of Freeport Sulfur until 1958, when he was replaced by Charles A. Wight (~1899 - 1972). Later heads of the business included Robert Chadwick Hills (1908 - 1998) in 1961, Thomas R. Vaughan (1908 - 1979) in 1968, Richard C. Wells (~1903 - 1986) in 1969, and Paul W. Douglas (1926 -) in 1975. Douglas had been president of Sulexco and was the son of Senator Paul H. Douglas (1892 - 1976). When Douglas was president, the chairman of Freeport Sulfur was Benno Charles Schmidt (1913 - 1999), the managing partner of the J.H. Whitney & Company (founded in 1946). Schmidt and Whitney had formed the firm in 1946. Schmidt coined the term "venture capitalist," and J.H. Whitney and Company became the first venture capital firm. A later Freeport Sulfur vessel was christened the *Benno Schmidt*. After Douglas in 1983, Robert Foster became president followed by R.L. (Willie) Williams, and Roy A. Pickren Jr. in 1985. In 1980, sulfur production had dropped to 2.3 million tonnes and continued to fall throughout the decade, down to 1.0 million tonnes in 1991, as reserves declined.

In 1981, Freeport Sulfur had merged with McMoRan to become Freeport McMoRan. Under the control of McMoRan, the sulfur business underwent various restructuring:

1986 – the sulfur and fertilizer assets of the company were transferred to Freeport-McMoRan Resource Partners;

1995 – on January 3rd, Freeport Sulfur acquired the assets of Pennzoil Sulfur (ex. Duval) including their Culberson mine.

1995 – Rio Tinto acquired a minority ownership in Freeport McMoRan.

1997 – in December of that year, Freeport McMoRan merged with IMC Global; the sulfur assets, with most of the Freeport McMoran management team, were spun off as an independent company named Freeport-McMoRan Sulfur.

1998 – late in the year, Freeport-McMoRan Sulfur merged with McMoRan Oil and Gas to become McMoRan Exploration Company.

The last, yet glorious, gasp of the Frasch industry was the opening of the Main Pass mine, 30 km off the coast of Louisiana by Freeport Sulfur in April 1991, at a cost of $880 million (initial estimates had been $550 million). This was the first new Frasch mine in over twenty years. The last Frasch mine in America was the second largest ever discovered in the U.S., with reserves of over 65 million tonnes. The project included fifteen offshore mining platforms and was over a mile long. Freeport, which had developed the only successful off-shore sulfur mines, was the major investor (58%) in Main Pass, which also included IMC (25%) and Homestake Mining (17%). The company had discovered sulfur there in 1988. The site was producing more than two million tonnes of sulfur per year by 1994. In September 1998, Hurricane George forced the evacuation of the off-shore rig at Main Pass. When the crews returned three days later, nine of the wells were sealed by solidified sulfur. Freeport had to purchase sulfur prills from Saudi Arabia and Poland to meet their sulfur commitments, because of the problems at Main Pass.

Though full capacity had been restored, the site was becoming less and less economically viable. Sulfur pricing was falling. The decision was made to close the mine. In 1999, the last full year of sulfur production, sales of the company were 2,973,500 tonnes, valued at $190 million (about 30% of the sulfur came from recovered sulfur sources). Of these volumes, 65% to 75% went to IMC, under a long-term contract, with the remainder mainly going to other phosphate producers. The giant, state-of-the-art Frasch complex at Main Pass was closed on August 31, 2000. Freeport purchased a shipment of 46,000 tonnes of sulfur prills from Vancouver for Galveston to meet their remaining contractual commitments for the latter part of 2000. The last president of Freeport was Robert M. Wohleber (1951 -). After the closure of Freeport, he became CFO of Kerr-McGee. Throughout the history of Freeport Sulfur (and associated companies), more than 165 million tonnes of sulfur were produced (a list of their Frasch mines appears in Table 4.3b).

In 2001, the company sold its sulfur lease at Main Pass to Trinity Storage Services but retained the rights to mine sulfur at the site. The Culberson mine property (from Duval) was sold for $3.2 million. The deserted mining platforms at Main Pass and Caminada were dismantled the following year. They still held prilling, remelting and sulfur storage facilities at Galveston, TX, Pensacola, FL and Tampa, FL. These were sold to Gulf Sulfur Services, a joint venture between IMC Global and Savage Service Corporation for $58 million (Freeport had initially been asking for $80 million). The McMoRan Exploration Company continues to operate today as an oil and gas developer in the Gulf. Savage Service Corporation operates the sulfur terminals to transport recovered sulfur from the Gulf to its partner IMC Global (now Mosaic) and other phosphate producers in Florida.

After a brief run of only nine years, Main Pass, the most expensive Frasch mine ever built, was shut down. With the closing of the last Frasch mine in the U.S., Joyce Ober of the U.S. Geological Survey wrote (**2000 Minerals Yearbook**):

After nearly a century of world dominance in the production of native sulfur, the U.S. Frasch industry shuttered its last mine as a result of low sulfur prices, continually increasing competition from low-cost recovered sulfur producers, escalating production costs, and technical problems.

Thus, the Frasch Century ended. In the one hundred years of the Frasch industry in the U.S., three hundred and thirty-five million tonnes of sulfur were produced!

Table 4.3a. Frasch company summary – Freeport Sulfur. Sales (MM US$) & production (1000 tonnes)

	1923	1924	1925	1926	1927	1928
Sales	6.1	4.9	7.2	9.4	13.4	13.2
Net income	0.7	-0.3	0.8	1.8	3.7	3.3
	1929	1930	1931	1932	1933	1934
Sales	14.8	13.9	102.	8.4	9.5	8.8
Net income	4.1	3.1	2.4	2	2.5	1.5
Production	1,403.6				285.3	414.1	450.9
	1935	1936	1937	1938	1939	1940
Sales	9.8	12.0	14.0	10.1	9.9	12.5
Net income	1.5	2.0	2.7	1.5	2.2	3.0
Production	612.8	598.9	711.5	684.2	791.0	860.4
	1941	1942	1943	1944	1945	1946
Sales	16	15.5	16.6	18.0	20.9	19.9
Net income	3.1	2.4	2.5	2.6	3.4	3.8
Production	905.2	1,014.9	1,027.8	1,0034	1,178.5	1,289.5
	1947	1948	1949	1950	1951	1952
Sales	20.2	24.2	27.9	33.8	34.8	38.3
Net income	3.1	4.3	5.9	6.8	6.3	7.3
Production	1,252.3	1,369.6	1,476.2	1,559.3	1,555.7	2,100.0
	1973	1974	1975	1976	1977	1978
Sales	167.5	247.5	291.8	308.8	312.1	300.0
Net income	30.0	78.2	34.5	48.5	21.2	27.9
	1993	1994	1995	1996	1997	1998
Production	1,403	2,088	3,0580	2,900	2,900	2,847

Table 4.3a. Frasch company summary – Freeport Sulfur. Sales (MM US$) & production (1000 tonnes)

Average price per tonne	57	53	70	62	61	62
	1999	**2000**	**2001**	**2002**		
Production	2,973	2,644	2,127	823		
Average price per tonne	63	54	34	37		

Sources: Anon., *Barron's,* (Oct. 7)**15** (40), p. 16 (1935); Self, S.B., *Barron's,* (May 16) **19** (20), p. 9 (1938); **U.S.G.S. Minerals Yearbook**; McConnell, *Barron's,* **20** (19), p. 8 (1940); Anon. *Barron's National Business and Financial Weekly,* (Jan. 3), **24** (1), p. 18 (1944); Wilder, T., *Barron's National Business and Financial Weekly,* (April 19), **34** (16), p. 17 (1954); Hussey, A.F., *Barron's National Business and Financial Weekly,* (Jan. 22), **59** (4), p. 24 (1979); company 10-K reports

Table 4.3b. Frasch company summary – Freeport Sulfur

Frasch Mines	**Opened**	**Closed**
Bryanmound	November 12, 1912	September 30, 1935
Hoskins Mound	March 31, 1923	May 26, 1955
Grand Ecaille	December 8, 1933	December 1978
Bay Ste. Elaine	November 19, 1952	December 1959
Garden Island	November 19, 1953	1991
Nash	February 5, 1954	November 21, 1956
Chacahoula	February 25, 1955	September 1962
Grand Isle	June 6, 1960	1991
Lake Peito	November 1960	1975
Caminada	February 1968	March 25, 1969
Caillou Island	October 1980	1984
Caminada (reopened)	October 1988	January 15, 1994
Main Pass	April 1992	August 31, 2000

4.3.2 TEXAS GULF SULFUR

As the Gulf Development Company was forming to exploit the Byranmound dome, another group from St. Louis had the same idea for Big Hill. Not only did they have similar ideas, they chose similar names for their companies; on December 23, 1909, the St. Louis group formed the Gulf Sulfur Company. Both "Gulf" companies evolved into the two greatest sulfur companies that have ever existed. Later, the Gulf saga continued as yet another Gulf company, the Gulf Production Company, became the largest shareholder in the Gulf Sulfur Company. Further confusing the history of the sulfur industry, a different Gulf Sulfur Company (see below) was formed by a Louisiana group to develop Frasch mines in Mexico.

Members of this St. Louis investment group were Alfred C. Einstein (~1866 - 1916), who was an electrical engineer and General Manager of Union Electric, John W. Harrison, J. M. Allen, and Theodore F. Meyer, who was president of the Meyer Brothers Drug Company [founded by Christian Frederick Gottlieb Meyer (1830 - 1905) in 1852]. The St. Louis sulfur syndicate, too, was dealing with Bernard Baruch. The big difference between the two start-up sulfur firms was that Gulf Sulfur could not find a financial backer like Swenson. While Union Sulfur had proven the technical and financial feasibility of such projects, the protracted litigation between Union Sulfur and Freeport Sulfur had deterred other investors from entering the business.

Finally, in late 1916, Bernard Baruch and J.P. Morgan, took control of the company. Seeley Mudd became the first president, but he resigned shortly afterwards to go into military service. He was succeeded by Walter Hull Aldridge (1867 - 1959). Aldridge had gotten his business experience in Canada. Back in 1898, the CPR had purchased the Trail, BC, smelter, and Aldridge had been made its manager; he then became a director of Cominco in 1904. In 1933, he received the William Lawrence Saunders Gold Medal, for distinguished achievement in mining, and, in 1950, the John Fritz Medal, the highest American award in the engineering profession, presented each year for scientific or industrial achievement in any field of pure or applied science. Aldridge remained president of Texas Gulf Sulfur until July 1951, when at the age of 83, he retired. Upon "retiring" from the president's office, Aldridge did not leave Texas Gulf Sulfur, but became its chairman. The new president was Fred M. Nelson (1897 - ?), who was in the office until early 1957.

On July 16, 1918, the name of the state was added to the name of the company, forming the Texas Gulf Sulfur Company. Less than a year later, on March 15, 1919, sulfur production began at Big Hill. Big Hill was a special site, for its geological formation was perfect for the Frasch process. The end result was that it was one of the most cost-effective sulfur domes ever developed, because of low hot water usage. The first rail car load of sulfur from the new Frasch mine was shipped to the sulfite pulp mill of Marathon Paper in Ontario. A major early customer had been Caesar Grasselli, who contracted for 68,000 tonnes of sulfur from the newest Frasch company. Texas Gulf Sulfur did their best to limit their disruptive influence on the market place, by focusing on the sulfuric acid industry, which was supplied by pyrites, mainly from Spain. Their competition was more Rio Tinto than Union Sulfur or Freeport Sulfur. The company also turned to Europe, where they hired Chance and Hunt to be their agents.

By 1925, Texas Gulf Sulfur controlled 40% of the domestic market. Their largest mine would not even open for four more years; in early 1929, Texas Gulf opened the Boling dome Frasch mine. While drilling for oil in the mid-1920's, the Gulf Production Company, a subsidiary of Gulf Oil, had repeatedly found sulfur in Boling Dome. In April 1927, core samples demonstrated that the deposit was huge. A bidding war erupted over the Texas treasure. Texas Gulf Sulfur, Freeport Sulfur and Union Sulfur all went after the rights to mine the property. Texas Gulf Sulfur beat out their rivals in the bidding war, but the price was high. Gulf Production received three million dollars and half of the net profits from the site, once the capital costs of the project had been recovered. However, by 1934, Texas Gulf Sulfur was still recovering its expenses. In that year, Gulf Production switched the future share of profits for 34% ownership of Texas Gulf Sulfur (1.3 million shares; in 1956, Gulf divested its shares). When finally the Boling Dome mine was closed in 1993, more than 80 million tonnes of sulfur had been recovered! About 15% of all the Frasch sulfur ever mined in the world came from this single mine, which operated for 64 years (most productive and longest operating Frasch mine in the world).

The sulfur industry, at this point, was relatively unknown to the general public; a publicity campaign was launched to promote the importance of this "exciting" industry. The yellow element hit the silver screen in 1941. The U.S. Bureau of Mines supervised the preparation of a motion picture called *The Story of Sulfur*. The movie was made in cooperation with Texas Gulf Sulfur and showed their state-of-the-art operations at Boling Dome. Another movie about sulfur, called *S Is For Sulfur*, was produced in 1952, to document the benefits of the Marshall Plan to stimulate sulfur production in Sicily after the Second World War.

In 1921, Texas Gulf Sulfur was listed on the New York Stock Exchange. Between 1922 and 1924, net income was $13.4 million, of which almost 90% was given back to shareholders as dividends. The stock price of Texas Gulf soared: $33 in 1921, reaching $122 in 1925 and peaking at $341 per share in 1929, before the Crash. By 1933, total dividends had reached over $80 million. During the Depression, Texas Gulf Sulfur never lost money. Their worst year was 1938, when their profit return on capital was 12%. Overall, between 1919 and 1943, profitability was high with an average profit-after-tax return of 22% on capital employed. Their chief rival, Freeport Sulfur, had a profitability of 14%, during the same 25-year period. Financial and other data during the Depression and later years are presented in Table 4.4a and the Frasch mines of Texas Gulf Sulfur are listed in Table 4.4b.

Between 1919 and 1943, Texas Gulf Sulfur produced a total of more than 28 million tonnes of sulfur, representing 58% of U.S. production. Boling Dome, and their other mines, continued to chug out sulfur and profits. In 1936, Texas Gulf produced more than one million tonnes of sulfur in a single year; other milestones followed: two million tonnes in 1942; three million tonnes in 1948. In 1948, Texas Gulf Sulfur produced twice the amount of sulfur as the other three producers combined. During the 1940's, the company began exploring for Frasch domes in Mexico, later opening up a mine (CEDI; see below). In 1959, Texas Gulf Sulfur entered the recovered sulfur business in Canada, with the opening of a sour gas plant in Okotoks, south of Calgary, and two years later, at Windfall, Alberta.

During the 1940's and 1950's, the company began diversifying into oil and gas, and other resource areas. In 1954, the company attempted to buy the Spanish pyrites assets of Rio Tinto, but this business went to local banking interests. In the early 1960's, the diversification program intensified. The company acquired phosphate rights and built a phosphate fertilizer plant in North Carolina; and a potash mine in Moab, Utah. Texas Gulf Sulfur expanded into other mining areas, where its major asset was the Kidd Creek zinc processing plant at Timmins, ON. In 1969, the company announced plans to build a $50 million zinc smelter at the site. The plant opened in April 1972.

The major mineral find at Timmins, in late 1963, led to one of the more infamous insider-trading scandals. Twelve directors and executives of Texas Gulf Sulfur were charged, including the president Claude O. Stephens (1908 - 1987), who had replaced Nelson, and executive vice-president, Dr. Charles Franklin Fogarty (1921 - 1981). The SEC took the company to court, and a verdict was issued on August 13, 1968. All but two of the twelve, including Stephens and Fogarty, were acquitted. Subsequently, Fogarty became president (1968) and later CEO (1973). In December 1971, the company agreed to pay $2.7 million to shareholders who sold their stock during the controversial period. In April 1972, Texas Gulf Sulfur became Texasgulf.

The Timmins asset and others in the north had made Texas Gulf Sulfur more a Canadian than American company, with over half of the corporation's income originating in Canada. In July 1973, the Canada Development Corporation (CDC) led a hostile takeover of Texasgulf, still under the direction of Fogarty, gaining 30% ownership in the company.

Tragically, on February 11, 1981, Fogarty, along with five other Texasgulf executives, died in a plane crash near White Plains, NY. The corporate jet was returning from a meeting with the CDC in Toronto. Five months later, in a bizarre move under the circumstances, the company was taken over by Elf Aquitaine (now Total) of France for $5 billion. On June 26, 1981, 75% of Aquitaine Canada (and later in the year Texasgulf gas operations in Canada and the Kidd Creek Mines) were sold off to CDC. The oil and gas assets were combined by CDC into a new company, Canterra (purchased by Husky in 1988), which became the largest sulfur producer in Canada and had the largest sulfur inventories in the world.

Elf Aquitaine kept control of the Frasch assets of Texasgulf in the U.S. The sulfur business that was once Texas Gulf Sulfur ended in 1993, when the legendary Boling mine closed. Later, in 1995, the remaining businesses of Texasgulf, which included the Aurora, NC, phosphate mine, were sold to the Potash Corporation of Saskatchewan (PCS) for $833 million.

Table 4.4a. Frasch company summary – Texas Gulf Sulfur. Sales (MM US$) & production (1000 tonnes)

	1919	1920	1921	1922	1923	1924
Net income*	1.0	3.3	1.9	3.9	4.7	4.8
	1925	**1926**	**1927**	**1928**	**1929**	**1930**
Sales					29.9	25.8
Net income*	5.7	9.4	12.1	14.5	16.2	14.0
	1931	**1932**	**1933**	**1934**	**1935**	**1936**
Sales	18.3	13.5	17.8	16.7	17.8	22.1
Net income*	8.9	5.9	7.4	7.0	7.5	9.9
Production		568.0	648.6	845.4	923.2	1,286.3
	1937	**1938**	**1939**	**1940**	**1941**	**1942**
Sales	26.0	17.4	20.9	25.6	29.8	28.8
Net income*	11.6	7.0	7.8	9.1	9.0	8.8
Production	1,723.8	1,222.6	818.3	1,428.4	1,839.4	2,173.4
	1945	**1946**	**1947**	**1948**	**1949**	**1950**
Production	2,203.4	2,091.0	2,67.3	3,120.0	2,797.4	3,153.6
	1951	**1952**	**1953**	**1960**	**1964**	**1965**
Sales		73.3	78.3	70.0	70.4	99.0
Net income*		25.1	24.5			
Production	3,146.8	3,200.0			2,400.0	2,600.0

* before depreciation from 1921 to 1932; after depreciation from 1933 to 1937
Sources: Anon., *Barron's*, (Feb. 28), **7** (9), p. 8 (1927); Self, S.B., *Barron's* (May 16), **19** (20), p. 9 (1938); **U.S.G.S. Minerals Yearbook**; Hussey, A.F., Anon., *Barron's*, (July 30), **14** (31), p. 12 (1934); Lawrence, H., *Barron's* (Nov. 4), **20** (45), p. 13 (1940); Anon. *Barron's National Business and Financial Weekly,* (Jan. 3), **24** (1), p. 18 (1944);

Table 4.4b. Frasch company results – Texas Gulf Sulfur

Frasch Mines	Opened	Closed
Big Hill (Gulf)	March 19, 1919	August 10, 1936
Boling	March 20, 1929	December 1993
Long Point	March 19, 1930	October 19, 1938
Clemens	May 3, 1937	April 18, 1949
Moss Bluff	June 24, 1948	1982
Spindletop	May 12, 1952	1976
Nopalapa (Mexico)	February 8, 1957	1959
Fannett	May 16, 1958	February 1977
Big Hill (reopened)	October 1965	December 1970
Nopalapa (reopened)	October 23, 1967	1959
Bully Camp	1968	July 1978
Comanche Creek	1975	November 1983
Comanche Creek (reopened)	December 1988	October 1989

4.3.3 DUVAL TEXAS SULFUR

During the 1920's, sulfur had been discovered at Palangana. Rights were offered to Union Sulfur and Texas Gulf Sulfur but they were not interested in the small deposit. A new company, called Duval Texas Sulfur, was formed to exploit the find. The founders were James Walker Cain and Alfred H. Smith. A financier of the company was Moody, Seagraves & Company. In 1930, they sold their interests to another of their companies, United Gas Corporation, who became the major investor (75% ownership). United Gas had been formed by Odie Seagraves and William Lewis Moody III in 1928, and the former was its president. United Gas placed George F. Zoffman (~1880 - 1957) in charge of the business. Sulfur production at the small dome began on October 27, 1928. Production during the Depression through after the War is presented in Table 4.5a and the major Frasch mines of Duval Texas Sulfur are listed in Table 4.5c.

Duval Texas Sulfur diversified into potash production, which became its major business. In 1950, the name of the company was changed to Duval Sulfur & Potash to reflect the changing portfolio. In 1959, it also began producing copper and molybdenum. In 1953, United Gas and Pennzoil merged, and the semi-independent business became Duval Corporation ten years later. The Frasch mine at Culberson, TX, opened by Duval in 1969, was the second last Frasch mine to close (then owned by Freeport) in the U.S. in 1999. The sulfur removed from this single mine was valued at over $600 million. The name Duval disappeared in

1985, when the group was merged into Pennzoil, becoming Pennzoil Sulfur (see Table 4.5b). During the late 1980's, Pennzoil Sulfur was briefly the largest Frasch producer in the world. In October 1994, Freeport Sulfur agreed to acquire the assets of Pennzoil Sulfur. The deal closed on January 3, 1995.

Table 4.5a. Frasch company summary – Duval Texas Sulfur. Production from 1932 to 1951 in tonnes

1932	1933	1934	1935	1936	1937	1938
22,599	37,971	44,695	64,347	118,712	132,032	212,406
1939	**1940**	**1941**	**1942**	**1943**	**1944**	**1945**
271,131	217,458	201,482	190,873	174,780	188,130	181,470
1946	**1947**	**1948**	**1949**	**1950**	**1951**	
253,165	176,460	167,530	220,855	201,475	220,000	

Source: **U.S.G.S. Minerals Yearbook**

Table 4.5b. Frasch company summary – Duval Texas Sulfur, Pennzoil Sulfur from 1989 to 1994

	1989	1990	1991	1992	1993	1994
Production (1000 tonnes)	1,968	2,128	1,696	1,171	655	n.a.
Sales (1000 tonnes)	2,186	2,091	1,970	1,690	1,071	n.a.
Revenue (MM US$)	n.a.	n.a.	225.5	148.0	74.0	71.9
Price* (US$ per tonne)	124	120	115	86	64	n.a.
Operation income (US$ MM)	n.a.	n.a.	42.7	1.0	-20.8	-57.4

(*) Average Green Markets Tampa Recovered Contract
Source: company 10-K reports

Table 4.5c. Frasch Company Summary – Duval Texas Sulfur

Frasch Mines	Opened	Closed
Palangana	October 27, 1928	March 10, 1935
Boling	March 23, 1935	April 25, 1940
Orchard	January 29, 1938	August 1970
Fort Stockton	1967	April 1970

Table 4.5c. Frasch Company Summary – Duval Texas Sulfur

Frasch Mines	Opened	Closed
Culberson	October 1969	October 30, 1999
Phillips Ranch	1980	1982

4.3.4 JEFFERSON LAKE SULFUR

Jefferson Island, LA, was named after the comic actor Joseph Jefferson (1829 - 1905), who was famous for his portrayal on stage of Rip Van Winkle. Here, he had a summer house. In 1895, while digging a well, he hit rock salt (i.e., a salt dome). Before becoming a legendary wildcatter, Captain Lucas had operated salt mines in this region while working for the Gulf Salt Company at Belle Isle in 1896 and for the Myles Salt Company in 1898. While with Myles, Lucas drilled test wells on Jefferson's property. Jefferson was offered $1 million for the property, but turned it down! Later, the Jefferson Island Salt Mining Company did extract salt on the island.

While the name "Gulf" confused the situation of the early history of Freeport Sulfur and Texas Gulf Sulfur, a similar situation arose between Duval Texas Sulfur and Jefferson Lake Sulfur. The word causing the confusion this time was the corporate use of "United." In 1930, the United Gas Corporation had become the largest investor in Duval Sulfur; five years earlier, the unrelated United Oil & Gas Syndicate had been formed and would eventually become known as Jefferson Lake Sulfur!

On April 30, 1925, the United Oil & Gas Syndicate was formed to explore for oil on Jefferson Island. The syndicate, founded by Arthur Barba (1861 - 1936) and under the leadership of Frank Coleman and then R.J.B. Abshire, had started as an oil exploration company. The company became Jefferson Oil & Development Company on November 10, 1926 and then the Jefferson Lake Oil Company on July 24, 1928. The firm had spent one million dollars looking unsuccessfully for oil. Approaching bankruptcy, the company, now under the leadership of A.A. Mayer, found sulfur in the center of Lake Peigneur. Jefferson Lake Sulfur wisely decided to switch their focus from oil to sulfur. The first sulfur production from Lake Peigneur was on October 18, 1932. The business officially became Jefferson Lake Sulfur Company on February 17, 1940. The president was now Joseph Mullen, and Adolphe D'Aquin was chairman.

From the late 1940's until 1964, the president of the company was Eugene H. Walet Jr. (1901 - 1968). In 1955, Jefferson Lake Sulfur and Socony Mobil Oil opened a sour gas plant at Manderson, WY. By the later 1950's, Jefferson Lake had also entered the recovered sulfur business in Canada. They were partners with Mobil Oil in the sour gas plant in Balzac, AB (outside of Calgary). The last Frasch mine opened by Jefferson Lake (Long Point, TX) was closed in 1982. Production during the Depression is presented in Table 4.6a and the major Frasch mines of Duval Texas Sulfur are listed in Table 4.6b.

In 1946, Jefferson Lake entered the carbon black business (the unprofitable venture was closed on August 31, 1948), and in 1956, they re-established an oil and gas division. They also entered the asbestos market. Occidental Petroleum purchased the company in 1964. In 1971, the Canadian oil and gas assets of Jefferson Lake Petrochemicals and Occidental Petroleum were merged to form Canadian Occidental Petroleum (Nexen).

Table 4.6a. Frasch company summary – Jefferson Lake Sulfur. Production from 1932 to 1951 (tonnes)

1932	1933	1934	1935	1936	1937	1938
13,401	303,787	76,135	24,473	8,439	94,374	268,129
1939	**1940**	**1941**	**1942**	**1943**	**1944**	**1945**
207,976	219,669	184,278	76,584	149,294	184,465	189,826
1946	**1947**	**1948**	**1949**	**1950**	**1951**	
226,021	245,180	212,118	250,566	277,837	355,723	

Source: **U.S.G.S. Minerals Yearbook**

Table 4.6b. Frasch company summary – Jefferson Lake Sulfur

Frasch Mines	Opened	Closed
Lake Peigneur	October 20, 1932	June 7, 1936
Clemens	May 3, 1937	December 1960
Long Point	June 7, 1946	1982
Starks	June 15, 1951	December 1960
Lake Hermitage	1968	March 1972

4.3.5 SULEXCO

Sulexco was not a producing company; yet it was a more influential factor in the world sulfur industry than any single producer. This export consortium of U.S. producers was the global sulfur-trading corporation. On October 4, 1922, the U.S. Frasch producers formed the Sulfur Export Corporation (Sulexco) to control the marketing of American sulfur overseas, especially in Europe. The original ownership of this company was Union Sulfur 37.5%, Freeport Sulfur 31.3% and Texas Gulf Sulfur 31.3%. Clarence A. Snider became the group's first president, a position he held for more than a decade. Snider had been associated with Frasch at the Frasch Process Soda Company from 1891 to 1898, where he had been treasurer; in 1907, he had joined Union Sulfur. Other original members of Sulexco were S. Magnus (VP); James Kilbreth (Secretary), and Charles Kemmler (Treasurer). The members of the Executive Committee of Sulexco were the presidents of the three sulfur firms, namely Henry Whiton of Union Sulfur, Eric Swenson of Freeport Sulfur and Walter Aldridge of Texas Gulf Sulfur. Their first offices were at 19-21 Dover Green, Dover, DE, moving in 1924 to 33 Rector Street, in New York City. These latter offices were at the corner of Rector Street and West Street in the 15-storey "Frasch Building", which had been erected in 1921 (the name and the building still exists; it was converted into condominiums in 2000). In 1926, after Union Sulfur involuntarily withdrew from the sulfur business, Sulexco was equally owned by the remaining two members. Three years later, the offices had moved again to 420 Lexington Ave, New York City, where they remained until 1952.

The sulfur corporation was an effective manager, some may say dictator, of global sulfur trade. Sulexco would control market share and pricing, outside of North America, for decades. Just after the formation of the organization, an agreement was reached with the Italians (COISS) in November 1922. Sulexco was given control of 75% of the global markets (outside of Italy and North America), with COISS controlling the remaining 25%. The global sulfur market was relatively simple at the time, with Sicily being the only serious export competitor and even then, the Italians were a distant second. The only other countries exporting sulfur were Japan, Chile and Spain, with their combined exports being a paltry 50,000 tonnes. During the Depression, the largest foreign markets of U.S. sulfur were Canada (excluded from the deal), followed by France, Germany and the UK in Europe, and Australia.

When COISS was replaced with UVZI, Snider wasted no time reaching an agreement on August 1, 1934, with the latest Italian consortium (see Appendix I). The deal divided the global market, excluding Italy and its colonies, North America and Cuba, between Sulexco and UVZI. The first 480,000 tonnes of sales were divided equally. Between 480,000 tonnes and 625,000 tonnes, Sulexco received 75%, plus 5,000 tonnes. Any global sales over 625,000 tonnes, 90% went to Sulexco. Even though Duval and Jefferson Lake were not part of the consortium, their volumes were included in the quotas of Sulexco. On November 18, 1935, the deal was officially suspended because the League of Nations declared sanctions against Italy for their invasion of Ethiopia. Sulexco and UVZI had already agreed on October 15th that their agreement would come back into force once sanctions were lifted. A clandestine, swap agreement (FSA-V) was reached whereby Sulexco would supply the countries with sanctions against Italy, while UVZI would supply all the needs of Germany, Austria and Hungary, and would continue to supply Switzerland, Argentina, and Brazil. By this surreptitious deal, UVZI, with the aid of Sulexco, side-stepped the restrictions of the sanctions imposed by the League of Nations. World War II put an end to the agreements between Sulexco and Sicily.

Sulexco was the "big brother" watching over the sulfur world. While they kept their "friends" in Italy under control, they kept a closer watch on the pyrites industry. In a bold move on the part of the latter, some of the larger pyrites giants started to produce elemental sulfur themselves. ORKLA brought on their technology to produce sulfur from pyrites, and built a world-scale facility in Norway. Rio Tinto had then opened a plant using the technology. The Spaniards tried to market their sulfur in neighboring France. ORKLA and Rio Tinto had crashed Sulexco's private party. The sulfur corporation again swung into action, first taking care of Rio Tinto. According to the FTC, Sulexco "exerted their influence to prevent such invasion of other markets." Further, on April 1, 1937, ORKLA and Sulexco reached an agreement to set pricing and market share, in Scandinavia and the Baltic States (see Appendix I). With this agreement, ORKLA would no longer license its technology to third parties. By this time, Norway had become the fourth largest producer of elemental sulfur in the world, behind Japan. Sulexco had swiftly neutralized the new competitors, with little damage to their position. The U.S. still accounted for more than 75% of global production.

The blatant cartel-like activity of Sulexco had been legalized in the U.S. by the Webb-Pomerene Act (to counter purchasing cartels). The major restriction of the legislation is that the actions of such cartels could not control market share or pricing in the U.S. The more successful and aggressive these organizations became, the more they naturally pushed the boundaries of the regulations. Sulexco had started to push too far. On March 14, 1939, the U.S. government began investigating the sulfur industry, especially Sulexco. In a related area, an indictment of anti-trust was brought down against fifteen sulfuric acid companies on June 26, 1942. Both cases were suspended until the war ended. On June 16, 1947, the FTC issued its final report entitled, The Sulfur Industry and International Cartels. The FTC objected to several of the agreements made with UVZI and ORKLA. Earlier on February 7th, the Commission had issued the following "recommendations" that Sulexco would have to "refrain from the future" from agreements that contained (see Appendix II):

    I.   U.S. export quotas, which incorporated the sales of non-members of Sulexco.

    II.   Minimal volume export guarantees to foreign competitors such as had been given to UVZI.

    III.   "Status quo" agreements to discourage competition in global markets, as had been established with UVZI.

IV.   Licensing agreements to restrict sulfur production technology transfer, such as had been reached with ORKLA.

V.   Marketing agreements with U.S. non-Sulexco members.

In summary, the FTC chastised Sulexco for virtually eliminating international competition through its trade agreements. The Commission warned Sulexco to not restrict trade in violation of the law. The mildly-worded FTC condemnation had been relatively weak, and there were no direct penalties. There was little direct impact on Sulexco at the time, especially because the agreements with EZI and ORKLA had not been in force since the start of World War II.

There was a major change in the strategy of Sulexco after the FTC edict. The FTC reprimand was more a warning, which Sulexco heeded, but the outcome was not what the FTC was expecting. What the FTC did not appreciate was that only these agreements kept the Italians in the global marketplace. Without them, the market would default to Sulexco because of their overwhelming competitive dominance. Sulexco had directly controlled more than 75% of world trade and had indirect control of another 20% or so, through their trade agreements. Since the deals were no longer valid, they went after the remaining 20% directly! Sulexco was no longer willing to "accommodate" their European competitors, especially the Italians. EZI had approached Sulexco but the latter refused to become involved in any discussions. Without the tacit support of Sulexco, EZI was doomed.

At the end of WWII, U.S. exports of sulfur soared (Figure 4.6). This growth was driven by increasing global demand, and problems with European supply of sulfur and pyrites. Sulexco had long dominated global sulfur markets, now their position grew even stronger. In 1948, the U.S. market share had increased to an incredible 93%! Sulexco had taken so much market share that the global market was teetering on the brink of disaster, as the U.S. soon found itself short of sulfur. The government had to step in to restrict exports (and domestic sales) for the sake of national security.

At the end of March 1952, Sulexco was dissolved by agreement between Freeport Sulfur and Texas Gulf Sulfur. Their own success had been their downfall. At the time, Sulexco had very little to do. Export and domestic allocations were under the control of the U.S. government. The powerful global trading firm had been relegated to an "unmarketing" organization, only deciding on whom not to sell to!

Afterwards, the sulfur world changed, much to the chagrin of the American Frasch companies. Suddenly, in 1955, there were 24,000 tonnes of sulfur imports into the U.S., which were almost zero before. A few years later, U.S. imports had escalated to 500,000 tonnes, and there was loss of market share in their overseas

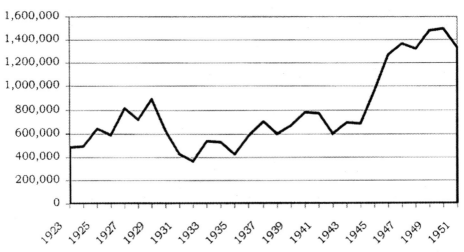

Figure 4.6. Sulfur - U.S. exports from1923 to 1951 in tonnes.

business as well. Frasch sulfur from Mexico began to be shipped to Florida and recovered sulfur, especially from Canada, was popping up everywhere. The U.S. now had some real competition.

Sulexco reformed in June 1958. The new group included Freeport Sulfur, Texas Gulf Sulfur, Jefferson Lake Sulfur and Duval Texas Sulfur, and handled all exports outside of North America. Their new offices were situated at 375 Park Avenue in New York City. The first president of the new Sulexco was Peter Black of Freeport Sulfur, until 1962 when Paul W. Douglas of Freeport Sulfur took the job. He was followed by Ernest A. Graupner (1908 - 1981), who had worked for Texas Gulf Sulfur, in 1966 until 1973.

What a change had taken place in the sulfur world during the 1950's. Sulexco started the decade with almost it all. By the end of the decade, the writing was on the wall. As Sulexco reformed, Mexican and Canadian suppliers were gnawing away their market share and major new competition came on stream from Lacq in France and a monstrous new native mine in Poland; for example, in 1960, non-U.S. sulfur production included:

APSA (Mexico)	1.3 million tonnes
SNPA (France)	0.8 million tonnes
Canada	0.5 million tonnes

Less than a decade earlier, elemental sulfur production from these three countries was zero. However, Sulexco shipments actually edged up in the 1950's, as global demand had grown. While sales had risen modestly, their market share was deteriorating.

There was only so much that they could do under the circumstances. The glory days were over. The situation hardly seemed fair, as more and more new sulfur poured onto the world scene every single year. The Mexicans, the French, the Polish, the Canadians, and the list kept getting longer; all had sulfur that cost almost nothing to produce. The Frasch industry was not so fortunate. Now they knew how the stunned Sicilian industry felt at the beginning of the century, and the same destiny lay ahead for them as well.

4.4 LIQUID SULFUR

In the 1960's, Sulexco had bounced back. This resurgence was led by their liquid sulfur logistics network. In 1912, Freeport Sulfur in the U.S. had developed the first liquid sulfur pipeline, which was encased in a metal steam jacket (i.e., a sulfur pipeline within a steam pipeline). The pipeline was used to transport sulfur from their Frasch well to the storage block. In these early days, liquid sulfur movements only took place within the confines of the sulfur mine site itself. All shipments to customers were broken sulfur. A milestone in the sulfur industry took place in 1947, when Freeport Sulfur transported liquid sulfur by tank car. By the 1960's, the rail fleet of Texas Gulf Sulfur had grown to more than 2,000 cars.

On July 31, 1952, the first insulated barge was launched. The barge was built to supply 900-tonne shipments to Consolidated Chemical Industries of Houston. On May 4, 1955, the first heated barges were introduced, permitting shipping greater distances. Generally, liquid sulfur is kept near 140°C, well above the melting point. A problem arises if it becomes too hot, >150°C, as plastic sulfur starts to form. The first heated-barge movement went from Freeport, at Port Sulfur, LA, to the National Lead Company, in St. Louis, MO; an eight-day journey of over 1,000 miles.

During the latter '50's, liquid sulfur became the dominant method to transport sulfur by the U.S. Frasch industry. In 1961, U.S. Bureau of Mines began the **Minerals Yearbook** with:

> The most notable event that occurred during the year was the conversion, by a large segment of the sulfur-consuming industry, to the use of molten sulfur. Estimates indicated that 50 percent of the elemental sulfur delivered to the customer was in molten form.

By 1962, 80% of the sulfur movements in the U.S. were in liquid form.

A sulfur terminal network was quickly established. In 1959, Texas Gulf Sulfur began construction of a liquid terminal at Beaumont, TX (opened the following year) which included 27,000 tonnes of liquid storage, and a storage terminal in Tampa, FL (opened February 1960). Texas Gulf Sulfur and Freeport Sulfur opened other terminals along the Eastern Seaboard and the Mississippi-Ohio Rivers. By 1965, there were already close to fifty storage tanks owned by sulfur producers scattered across the U.S., with a total capacity of more than 450,000 tonnes. Another 165,000 tonnes of storage capacity were located at major customers.

After rail and barge, the next natural development was vessel movements. Texas Gulf Sulfur initiated the first movement of liquid sulfur by vessel in 1961, with the *Sulfur Queen* (ex. *Esso New Haven*). Also brought into service was the *Louisiana Sulfur* (ex. *Robert E. Hopkins*) by Freeport Sulfur in the same year. The former vessel became famous for the tragedy that struck her. As with most of the early liquid sulfur vessels, the *Sulfur Queen* was a World War II-vintage, T2 oil tanker. The ship had been built in March 1944, by Sun Shipbuilding of Chester, PA. The vessel was converted for liquid sulfur by Bethlehem Steel for Texas Gulf sulfur in 1960. Its first sulfur load was made on January 1961. During its first year, the *Sulfur Queen* moved 434,000 tonnes of molten sulfur. The ship was on one of its routine shuttle voyages from Beaumont, TX, to Norfolk, VA, with 15,000 tonnes of sulfur, when the ship disappeared off the Florida coast. The last message from the *Sulfur Queen* was sent at 1:25 AM on February 4, 1963. The ship was never heard from again. Thirty-nine crew members perished. The lost vessel found sudden fame in the 1975 book by Charles Berlitz, **The Bermuda Triangle**.

Soon liquid sulfur was being shipped across the Atlantic Ocean. In 1963, Sulexco entered into a long-term lease agreement for a new liquid sulfur terminal in Rotterdam. The following year, liquid sulfur was shipped from Beaumont, TX, to Rotterdam, using the *Naess Texas*. This vessel, constructed at Haverton Hill, England in the yards of the Furness Shipbuilding Company, was the largest ship of its kind. The vessel was commissioned for Sulfur Carriers Inc., with a long-term charter to Sulexco, and operated by Naess Denholm. The ship was equipped with four tanks, which were heated by steam coils, using waste heat from the ships engines. Its sister ship, the *Naess Louisiana*, entered service late in the same year. By the end of 1964, eight ships were in trans-Atlantic sulfur service. Sulexco sales hit new records in 1965. By the late 1960's, over one-third of the group's sales were through the Rotterdam terminal. Later, Antwerp replaced Rotterdam as the major European port of entry for American sulfur.

Liquid sulfur was becoming the state of choice for global sulfur customers. The U.S. Bureau of Mines reported (**Minerals Yearbook** 1973):

> There is a continuing trend towards the use of liquid sulfur tank ships in international trade and the installation of liquid sulfur terminal facilities at points of consumption. This was being brought about because of environmental problems associ-

ated with the storage and shipment of dry bulk elemental sulfur with its associated dust problems, and the preference of consumers for the delivery of liquid sulfur...

By the mid-1990's, there were sixteen liquid sulfur tankers in service, including the state-of-the-art *Sulfur Enterprise* of Freeport Sulfur. The *Sulfur Enterprise* is still in active service, under the Gulf Sulfur Service subsidiary of Savage Service Corporation. With the demise of the Frasch industry, the vessel now moves recovered, instead of Frasch, sulfur from Galveston, TX, to Tampa, FL. Another major vessel in active service is the *MV Aurora* (named after the phosphate plant in Aurora, NC). This ship is chartered by Potash Corporation of Saskatchewan (who had purchased the phosphate facility once owned by Texas Gulf Sulfur) for moving liquid sulfur from the Amuay oil refinery of Lagoven S.A. in Venezuela to North Carolina.

Trans-Atlantic movements fell into disfavor as Europe began obtaining most of its sulfur locally especially from France and later Poland. For example, the *Marine Duval* (scrapped in 2002) that had been carrying liquid sulfur across the Atlantic, was converted to Gulf service for Freeport Sulfur.

4.5 NON-U.S. FRASCH MINES

4.5.1 MEXICO

Native sulfur mining in Mexico began when the conquistadors arrived in the New World. In 1522, Francisco Montano, one of the officers of Cortez (1485 - 1547), was lowered into the crater of the active volcano Popocatepetl (70 km southeast of Mexico City). Cortez wrote of the crazy sulfur mining expedition to the Holy Roman Emperor Charles V (1500 - 1558; known as Carlos I in Spain):

> ...out of [the volcano] sulfur has been taken by a Spaniard who descended seventy or eighty fathoms by means of a rope attached to his body below his arms...It is hoped that it will not be necessary to resort [again] to this means of procuring it.

Enough sulfur was extracted to produce fifty caskets of gunpowder.

Between 1850 and 1865, a sulfur distillation plant was built on the side of the volcano, but various attempts to mine sulfur from the crater failed. The highest native sulfur production in Mexico was reached in 1912 at 9,600 tonnes. Most native sulfur was coming from the vicinity of Mexicali. During the 1930's, Jefferson Lake Sulfur had tested deposits from San Felipe, southeast of Tijuana, and again in the late 1940's, but no further developments were made. In 1940, the Colima and Ransburg Syndicate started developing a sulfur deposit at Cuidad Guzman, in south, central Mexico.

Even before the opening of the first Frasch mine in Mexico, recovered sulfur was produced. In 1950, Pemex (Petroleos Mexicanos S.A. de C.V.) opened the first recovered sulfur from sour gas in Mexico at Poza Rica. Production was 37,000 tonnes in 1952, the second largest recovered sulfur plant in the world at the time. While an early entrant into this field, expansion came slowly, unlike the developments in France and Canada. By 1961, recovered sulfur production had only reached 50,000 tonnes per year. Recovered production today is over one million tonnes.

As Pemex was preparing to enter the recovered sulfur market, developments were also taking place in Frasch mining. Sulfur deposits, such as those at Calcasieu, were first found in Mexico in 1904, by the Mexican Eagle Oil Company (Cia Mexicana de Petroleo El Aguila), founded by Weetman Dickenson Pearson (1856 - 1927), Lord Cowdray. In 1889, Pearson had been invited by President José de la Cruz Porfirio Diaz (1830 - 1915) to build a railroad in Mexico. During the construction, the company saw oil seeping out of the ground, and in 1900, Pearson formed Mexican Eagle Oil. In 1919, the firm was purchased by Shell. Mexican Eagle Oil went on to be the leading oil company in the country until expropriated by the government of Lázaro Cárdenas del Río (1895 - 1970) in 1938, and placed under the control of Pemex. Mexican Eagle Oil had been given full control of the Poza Rica field (discovered in 1932) only in the previous year.

However, Mexican Eagle Oil had been looking for oil, and the sulfur find was buried in the company files. The Mexican Frasch industry was opened up by three brothers from Louisiana: Lawrence Belser Brady (1895 - ?), a lawyer; Ashton G. Brady (1900 - 1978), a geologist; and William Brady (~1902 - ?), a specialist in real estate. They were the sons of Willy G. Brady (1858 - ?) and Caro Brady (1863 - ?) of Saint Mary, LA. The "Brady Sulfur Bunch" formed the American Sulfur Company. In 1940, Ashton Brady was going through the 1904 exploration report of the Mexican Eagle Company and saw the lost sulfur discovery. Two years later, they started drilling new exploratory wells in the same area. The first two wells were dry, but with the third they hit sulfur. After exploring swamps and jungles in the region two more major domes were found. They had first approached Texas Gulf Sulfur and Freeport Sulfur, but the sulfur giants showed little interest. They then approached various investment syndicates, which purchased control of the sulfur rights of the Brady's American Sulfur Company: San Cristobal to Mexican Gulf Sulfur; Jaltipan to Pan-American; and Salinas to Gulf Sulfur (Azufre Vera Cruz). The only Mexican Frasch company that did not originate from the exploration of the Brady Brothers was CEDI, the Texas Gulf Sulfur operations. The adventures of these Mexican sulfur explorers were presented in a *Time Magazine* article on February 21, 1955.

In 1946, the Mexican Gulf Sulfur Company (Mexsul) was organized to exploit the first of the Brady properties. The company was founded by Eugene Norton, who had been the interim president of the Freeport Sulfur Company before Williams (see above), Edwin D. Belknap, Edwin S. Gardner and Cyrus M. Lerner. Norton became the company's first president. Four wells were drilled in this year by Mexican Gulf Sulfur with the American Sulfur Company. However, many years passed, before a Frasch mine was opened. Norton was succeeded by Paul Nachtman in 1949 and then Robert H. Van Doren became president of the company. By the mid-1950's, John Tyler Claiborne Jr., a colleague of Norton back in the Freeport Sulfur days, was now running Mexican Gulf Sulfur. The sulfur crisis brought on by the Korean War led the company to make the major investment. On March 15, 1954, the U.S. Frasch monopoly ended with the opening of the first Frasch mines outside of the U.S. by Mexican Gulf Sulfur at San Cristobal. The company had been in development longer than it would be in operation. The initial Mexican investment in Frasch sulfur proved to be a failure. The sulfur reserves were limited and the company soon folded. During their brief period of operations, production was:

1954	55,000 tonnes
1955	83,675 tonnes
1956	13,425 tonnes

An auction was held on April 11, 1958 to sell off the company's assets. Having failed to attract the minimum bid, the Mexican government purchased the assets, through National Financiera, for $2.5 million.

The second Brady project at Jaltipan was the antithesis of San Cristobal. Unlike the poor reserve of Mexsul, this other deposit turned into one of the great sulfur domes. This option went to the Pan American Sulfur Company (PASCO) of Houston, through its Mexican subsidiary APSA (Azufrera Panamericana SA). PASCO was formed on April 17, 1947 by a group of Dallas businessmen, including the oil executive Jabal Richard Parten (1896 - 1992), the company's first president; his long-time partner Sylvester Dayson (1892 - 1978), Douglas W. Forbes, who followed Parten as president; C. Andrae III, Roland S. Bond (1898 - 1976), Elijah E. (Buddy) Fogelson (1900 - 1987), who two years later married the actress Greer Garson; and Guy I. Warren. Sulfur first flowed from Jaltipan on September 24, 1954. The growth of APSA came as a shock to its American rivals. A big coupe for the company was its first major U.S. client. In 1955, APSA signed a ten-year contract to supply 150,000 tonnes per year of sulfur to General Chemical. The mine reached a milestone on December 10, 1956, producing its one millionth tonne of sulfur, and it produced more than one million tonnes of sulfur in a single year in 1960. By 1958, APSA was the third largest sulfur producer in the world, after Texas Gulf Sulfur and Freeport Sulfur. A summary of the early years of Mexican Frasch sulfur exports, mainly from APSA, is presented in Table 4.7. By the mid-1960's, Harry C. Webb, who had worked with Texas Gulf Sulfur, was president of PASCO, and Marlin E. Sandlin (1909 - 1974) who had worked for Parten at the Woodley Petroleum Company was chairman.

Table 4.7. Sulfur - Mexico. Exports from 1954 to 1960 (tonnes)

	1954	1955	1956	1957	1958	1959	1960
U.S.	0	30,547	239,146	489,455	607,381	673,628	589,910
U.K.	0	28,821	48,063	80,997	58,822	71,608	89,038
France	0	11,507	11,507	105,267	144,696	77,820	61,932
Australia	0	42,460	40,606	58,066	72,592	46,636	74,196
Total	49	117,501	493,992	863,247	1,005,501	1,064,784	1,172,249

Source: **U.S.G.S. Minerals Yearbook**

The third dome at Salinas was, at first, kept by the Bradys, being placed in their Gulf Sulfur Company. The first sulfur flowed from Salinas on May 3, 1956. Just before this, the Bradys sold their share of the company to the New York investment house of Baer, Stearns & Company, and the Houston firm, Hudson Engineering Company. Upon the take-over, Stuart Callender Dorman (? - 1978) succeeded his father-in-law Lawrence Brady as president of Gulf Sulfur. The sulfur business struggled, not turning a profit until 1961, then under the direction of Robert H. Allen (1927 -), who had become president in 1957. When the government passed regulations forcing them to sell most of their production only in Mexico, Gulf Sulfur attempted to dispose of their Mexican assets. After the government blocked the sale of the company to the Mexican firm Inversiones Azufreras in 1969, the operations folded at the end of the year. Allen, though, continued to grow the company through non-sulfur investments in the U.S. By 1982, Allen had increased sales of the company to over $400 million. Gulf Sulfur became part of Gulf Resources and Chemical Corporation (now Gulf USA Corporation).

In 1949, Texas Gulf Sulfur reached a deal with the Mexican government to mine sulfur in the region of Nopalapa (Cia Exploradora del Istmo; CEDI). Finding commercial deposits is very difficult and somewhat haphazard. Not having the advantage of a dome already discovered for them, the first sulfur was not produced until February 7, 1957. CEDI was the last of the sulfur companies in Mexico to open and the last to close.

Mexican sulfur was exported from the port of Coatzocoalcos, near Vera Cruz. APSA had set up sulfur terminals in Tampa (1960) and Immingham, in England. In 1961, the liquid sulfur vessel *Etude* (ex. *Antelope Hills*), later replaced by *H.H.*

Jaquet (to U.S.; named after Harold Jaquet, the vice-president of operations for APSA), entered the fleet. Other early vessels were the *Pochteca* (ex. *Atlantic Refiner*) and the *Harry C. Webb*, which was later rechristened the *Otapan* (to Europe). Mexican shipments to Europe started in 1965. In October 1989, Pemex launched the *Teoatl*. These vessels are now out of service, and Pemex is chartering vessels from Polsteam.

A crisis broke out in the Mexican sulfur world on April 26, 1965. The government of Gustavo Diaz Ordaz (1911 - 1979) had abruptly passed legislation that established export quotas on sulfur. APSA was caught off guard; the loading of the *Surrey Trader*, destined to Australia for APSA with 16,000 tonnes of broken sulfur, was ordered stopped by the government; three other ships were also waiting to be loaded. The government had taken the first step to taking over the industry. In 1938, the oil industry had been nationalized, but the American-owned sulfur companies thought that a similar fate did not await them. They were wrong. The official reasoning for the government action was that the sulfur reserves of the country would be depleted in five years. The entire affair has many similarities to the TAC take-over of the early Sicilian industry.

For the American owners of these firms, especially the largest APSA, the situation did not look good. The quota was cut again the following year. In early 1966, APSA announced plans to join a consortium, including the Banco Nacional de Mexico, for the construction of a large fertilizer plant at Coatzocoalcos. The investment was forced upon the company, in the hope that the authorities would back off. The strategy did not work. On June 30, 1967, 66% of APSA was sold to Mexican interests for $50 million; the remainder of the company was purchased from PASCO for $10 million in 1972. PASCO was liquidated four years later; Webb later became president of Underwood, Neuhaus & Company in 1982 (the investment firm became Lovett, Underwood, Neuhaus & Webb and was later taken over by the Kemper Securities Group).

An agreement was also reached with Texas Gulf Sulfur and its subsidiary CEDI; the ownership was divided equally between Texas Gulf Sulfur, the Mexican government, and a Mexican investment group. As discussed above, Gulf Sulfur folded under the export restrictions. The Mexican government then controlled APSA and CEDI through the Commission for Mining Development (Comision de Fomento Minero, CFM) and other public entities:

APSA: CFM (55.33%) and Nacional Financiera S.N.C. (40.65%), plus
private investment from Banco Nacional de México S.N.C. (4%); and
Minera Carbonifera Rio Escondido (0.01%).
CEDI: CFM (51%) and Fertilizantes Mexicanos (13%), plus Texas Gulf Inc.
(34%), and two private Mexican concerns (2%).
All sulfur exports from Mexico were placed under APSA.

In 1971, the U.S. investigated dumping charges against Mexico, and duties were imposed the following year. Mexican production peaked in 1974 at 2.2 million tonnes, with most of the production coming from APSA. The Mexican fertilizer industry had greatly expanded in this period and was now consuming 500,000 tonnes of sulfur per year.

Recovered sulfur proved to be the downfall of the Mexican Frasch industry as well. In August 1992, APSA declared bankruptcy. The debt of the company was $220 million. APSA closed its three mines in November 1992, and CEDI closed its mine in May 1993. Total sulfur production from the Frasch industry in Mexico was 55 million tonnes (see Table 4.8, and Figure 4.7). The assets of APSA and control of Mexican sulfur exports were assigned to Pemex (becoming their Texistepec Mining Unit) by the Mexican government in lieu of prior sulfur sales owing. Sulfur continued to be produced from their oil refineries at Salina Cruz and Tula. Pemex operates nine sulfur recovery units, and produces over one million tonnes of recovered sulfur per year.

Table 4.8. Frasch company summary – Mexico sulfur production in 1000 tonnes

Company	1950	1955	1960	1965	1985
APSA	0	391	1,028	1,143	868
Mexican Gulf Sulfur	0	84	0	0	0
Comp. Azufre Veracruz (Gulf Sulfur)	0	0	203	339	0
CEDI (Texas Gulf Sulfur)	0	0	0	0	663
Central Minera (Texas Intern. Sulfur)	0	0	18	0	0
Total	0	475	1,249	1,481	1,551

Source: **U.S.G.S. Minerals Yearbook**

4.5.2 POLAND

The next major sulfur powerhouse was Poland. As Mexico was expanding into Frasch, some of the largest native (non-Frasch, open-pit) sulfur mines in the history of the industry were under development in Poland. The first was at Piaseczno in 1958, discovered by the geologist Pavlovski four years earlier, and the second, the world's largest native (open pit) sulfur mine, was at Machow (opened in January, 1964, closed 1993). Before any sulfur was removed from Machow, millions of tonnes of earth had to be removed. The first sulfur did not come from Machow until the end of 1969. Poland became the largest native (non-Frasch) sulfur producer in the history of the industry.

The Polish sulfur industry involved a number of state-owned enterprises:

- *Siarkopol* - includes two companies involved in sulfur production: Kopalnie i Zaklady Przetwórcze Siarki w Tarnobrzegu (Sulfur Mines & Reprocessing Works in Tarnobrzeg; including the Jeziorko "Siarkopol" Grebow-Wydza deposits; in 2001 became Zaklady Chemiczne, Siarkopol Tarnobrzeg), formed in 1954. Kopalnie i Zaklady Chemiczne Siarki w Grzybowie (Sulfur Mines and Chemical Works in Grzybów, including the Osiek and "Siarkopol" Grzybow-Gacki deposits), formed in 1967.
- *Siarkopol Gdansk* - opened a liquid sulfur terminal at Gdansk in 1971 - storage and transfer of sulfur from land to ships.
- *Polsteam* (Polska Zegluga Morska) - three liquid sulfur vessels were in service by 1973; mainly in shuttle service to Rotterdam - shipping of liquid sulfur by vessel
- *Ciech* - Centrala Importowo-Eksportowa Chemikalii i Aparatury Sp. (Import and Export of Chemicals and Chemical Equipment Enterprise Co), formed in 1945; from 1987 to 2001, Jozef Karolak (? - 2001) was director of the sulfur operations of Ciech; in 1993, Ciech Siarkopol opened two wet prilling plants in Gdansk, which had enough capacity to prill all Polish production. Today, Ciech S.A. has five operational divisions; sulfur is marketed through Ciech Agrosiarka - sulfur trading and marketing.

Figure 4.7. Frasch sulfur - Mexico production from 1953 to 1994 in tonnes.

In 1970, Polsteam had entered the sulfur tanker business with an agreement with Siarkopol; they launched their first vessel the *Norvest*, the following year, and in 1974, the *Professor K. Bohdanowicz* (9,000 tonnes), named after Karol Bohdanowicz (~1865 - 1947) who had been head of the National Geological Institute in Poland, was also placed into service. During the 1970's, Polsteam was one of the leading sulfur carriers in the world, transporting over 800,000 tonnes of sulfur per year. The Polsteam fleet of liquid sulfur tankers was modernized in the late 1990's with the addition of four new vessels: the *Penelope* (1996), *Kaliope* (1997), *Mitrope* (1999), and the *MV Aurora* (2000). In 2005, the Polsteam sulfur fleet had a cargo capacity of almost 50,000 tonnes. After the demise of the Polish Frasch industry, many of the vessels were contracted by MG Chemiehandel of Germany.

As Machow was under development, the Frasch process was introduced at other deposits. Poland was the third country to open a Frasch mine at Jeziorko (in Rzeszow) and Grzybow (in Kielce) in 1967. A modified Frasch process, called the hydrodynamic process, was developed by Polish engineers. The modifications allowed for a Frasch-type process to be used in deposits more similar to those of Sicily, as existed in Poland. In 1970, Poland surpassed Mexico as the second largest Frasch producer in the world, and then became the world's largest Frasch producer in 1982, surpassing that of the U.S. In 1977, production was 2.9 million tonnes at Jeziorko and 1.4 million tonnes at Grzybow. During this period, Poland was the only supplier to the communist-block countries, and they had 30% market share in Western Europe. Through much of the 1980's and the 1990's, Poland was the second largest sulfur trader, behind only Canada. Major contracts were signed with Morocco and Brazil. In September 1993, the Osiek Frasch mine, near Grzybow opened. Osiek has the distinction of being the newest major Frasch mine in the world. By the end of the decade, sulfur reserves were declining in the mines, as were exports (see Table 4.9). The Jeziorko mine was closed in 2001, leaving Osiek as the only operating Frasch mine in the world. Polish production has dropped below one million tonnes per year (see Figure 4.8). By 2005, Poland had produced over 100 million tonnes of Frasch sulfur.

Table 4.9. Sulfur – Poland. Exports in 1000 tonnes

	1997	1996
Austria	-	7
Belgium	-	7
France	66	56
Germany	19	3
Italy	53	63
Norway	3	-

Table 4.9. Sulfur – Poland. Exports in 1000 tonnes

	1997	1996
Sweden	90	88
UK	38	29
Egypt	-	28
Morocco	376	484
Tunisia	110	221
Indonesia	-	14
Brazil	210	180
Czech Republic	29	60
Croatia	35	25
Slovenia		5
Others	6	6
Total	**1,050**	**1,276**

Source: *Weekly Sulfur Report*, FMB Consultants

4.5.3 U.S.S.R.

In March 1971, a Frasch pilot plant was opened at Gaudak, Turkmenistan, with a full commercial unit starting up the following year. The Frasch operations were expanded in the mid-1970's. This unit supplemented the massive native sulfur industry, with huge mines at Rozdol and Gaudak (smaller native mines included Yavorov, Shorsu, Changyrtash, Alekseyev, Vodnin, and Zapadnyy).

4.5.4 IRAQ

In 1970, the mining company Hydrokop and the mining engineering institute Centrozap, both of Poland, had helped design the Frasch mine at Mishraq, 50 km south of Mosul. The mine sits on the largest native sulfur reserves in the world, which may surpass 200 million tonnes. In January 1972, the mine was opened at a cost of $54 million by the government-owned National Iraqi Minerals Company (NIMCO), but it is operated by the Mishraq Sulfur State Enterprise, out of Mosul. There is a sulfuric acid and aluminum sulfate plant at the site. In 1974, the mine produced 570,000 tonnes of sulfur, and had a capacity of one million tonnes. Iraqi sulfur was being exported to China, India, Lebanon and Poland.

Political instability has hampered Iraqi sulfur production for almost three decades. Beginning in September 1980, shipments were disrupted by the war with Iran. Before Operations Desert Storm in 1991, the mine was planning to expand capacity to two million tonnes per year. Through the period of sanctions against Iraq, the mine continued to produce, but at reduced rates. In late 2000, an agreement was made to supply all of Jordan's sulfur requirements, mainly to Jordan

Figure 4.8. Frasch sulfur — Poland production from 1966 to 2004 in tonnes.

Phosphate Mines. Despite the Jordanian agreement, customers were few and the mine continued to build inventories.

On June 24, 2003, a major fire took place at the mine. After burning for a month, 300,000 tonnes of sulfur inventory had been destroyed. The massive sulfur dioxide plume spread over 1,000 km, and was tracked by satellite. The 600,000 tonne gas cloud was the largest man-made release of sulfur dioxide in history. Hussain's troops had added to the problem by flooding the mines with bitumen.

In 2004, Ciech of Poland studied the possibility of reopening the Mishraq sulfur mine. However, the instability of the region forced them to delay any plans to reopen the site. Since the devastating fire of 2003, the mine is believed to have been closed.

4.5.5 BRAZIL

In 1983, a 300 tepd-pilot plant to produce Frasch sulfur at Castanhal, Sergipe, opened. The plant produced about 6,000 tonnes of sulfur per year. The last year data was reported by the U.S.G.S. in 1993.

Many sites have utilized the famous process of Herman Frasch (see Table 4.10). With opening of the Frasch mines in Poland, global Frasch production (including the U.S. and Mexico) surpassed the ten million tonne mark in 1969. Frasch production peaked in the mid-'70's at over 14 million tonnes. By the early 1990's, the global industry had started to collapse. The Frasch process did not fall

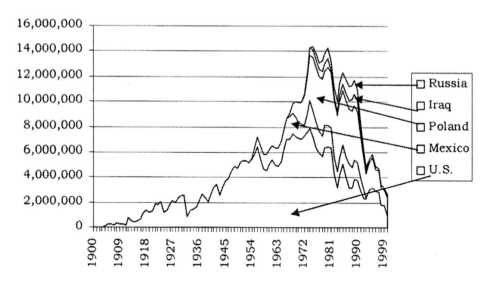

Figure 4.9. Frasch sulfur – World production from 1900 to 2000 in tonnes.

to a better technology, it fell to a better source. With recovered sulfur, the yellow element is essentially free. How could any process compete against such an economic deterrent? The Mexican industry closed down. Mishraq, as usual, was caught in the political turmoil of the region. Polish production was also scaled back because of serious economic and environmental problems, and declining reserves. The political fallout from the collapse of the U.S.S.R. hampered production in the Ukraine and Turkmenistan. Only the Main Pass mine kept U.S. production going. The demise of this grand site in 2000 marks the practical end of the Frasch industry. The business had run its course and has now essentially shut down. In the Frasch century, over five hundred million tonnes of elemental sulfur was produced by the technology of the perpetual entrepreneur from Gaildorf (see Figure 4.9).

Table 4.10a. Frasch mines – U.S. production in 1000 tonnes

Sulfur Domes	Operator	Opened	Closed	Production
Texas				
Big Creek	Union Sulfur	1925	1926	
Big Hill (Gulf) (3rd)	Texas Gulf Sulfur	1919	1932 (1936)	12,346
Bill Hill (reopened)	Texas Gulf Sulfur	1965	1970	

Table 4.10a. Frasch mines – U.S. production in 1000 tonnes

Sulfur Domes	Operator	Opened	Closed	Production
Boling (Newgulf) (6th)	Texas Gulf Sulfur	1929	1993	80,800
Boling	Union Sulfur	1928	1929	8
Boling	Baker-Williams	1935	1935	2
Boling (10th)	Duval Sulfur	1935	1940	571
Bryanmound (2nd)	Freeport Sulfur	1912	1935	5,001
Bryanmound	Hooker Chemical	1966	1969	
Clemens	Jefferson Lake Sulfur	1937	1960	3,000
Clemens	Texas Gulf Sulfur	1937	1949	746
Comanche Creek	Texas Gulf Sulfur	1975	1983	
Comanche Creek (reopened)	Texas Gulf Sulfur	1988	1989	
Culberson (Pecos)	Duval Sulfur	1969	1999	43,977
Damon Mound	Standard Sulfur	1953	1957	140
Fannett	Texas Gulf Sulfur	1958	1977	
Fort Stockton	Duval Sulfur	1967	1970	
Fort Stockton	Atlantic Richfield	1968	1985	
High Island	U.S. Sulfur	1960	1962	
High Island	Pan Amer. Petr. (Amoco)	1969	1971	
Hoskins Mound (4th)	Freeport Sulfur	1923	1955	10,865
Long Point (7th)	Texas Gulf Sulfur	1930	1938	402
Long Point	Jefferson Lake Sulfur	1946	1982	
Long Point	Admiral (Lone Star)	1956	1956	0
Moss Bluff	Texas Gulf Sulfur	1948	1982	
Nash	Freeport Sulfur	1954	1956	149
Nash (reopened)	Phelan Sulfur	1966	1969	
Orchard	Duval Sulfur	1938	1970	
Palangana (5th)	Duval Sulfur	1928	1935	238
Phillips Ranch	Duval Sulfur	1980	1982	
Spindletop	Texas Gulf Sulfur	1952	1976	
Louisiana				
Bay Ste. Elaine	Freeport Sulfur	1952	1959	1,131
Bully Camp	Texas Gulf Sulfur	1968	1978	
Caillou Island	Freeport Sulfur	1980	1984	
Caminada (*)	Freeport Sulfur	1968	1969	
Caminada (reopened)	Freeport Sulfur	1988	1994	
Chacahoula	Freeport Sulfur	1955	1962	

Table 4.10a. Frasch mines – U.S. production in 1000 tonnes

Sulfur Domes	Operator	Opened	Closed	Production
Chacahoula (reopened)	U.S. Oil (John W. Mecon)	1967	1970	
Garden Island Bay	Freeport Sulfur	1953	1991	
Grande Escaille (9th)	Freeport Sulfur	1933	1978	36,700
Grand Isle (*)	Freeport Sulfur	1960	1991	
Lake Hermitage	Jefferson Lake Sulfur	1968	1972	
Lake Peigneur (8th)	Jefferson Lake Sulfur	1932	1936	431
Lake Pelto	Freeport Sulfur	1960	1975	5,622
Main Pass (*)	Freeport Sulfur	1992	2000	18,000
Starks	Jefferson Lake Sulfur	1951	1960	900
Sulfur (1st)	Union Sulfur	1895	1924	9,400
Sulfur (reopened)	Union Texas Petroleum	1966	1970	

Table 4.10b. Frasch mines – non-U.S. production in 1000 tonnes

Sulfur Domes	Operator	Opened	Closed	Production
Brazil				
Castanhal		1983	?	small
Iraq				
Mishraq	Mishraq Sulfur State Enterprise	1972	2003	may reopen
Mexico				
Coachapa	APSA	1981	1992	
Jaltipan	APSA (Pan American Sulfur)	1954	1992	
Las Salinas (Amezquite)	Com. Azufre Veracruz (Gulf Sulfur)	1955	1969	
Nopalapa	Cia Explor. del Istmo (CEDI)	1957	1960	385
Nopalapa (reopened)	Cia Explor. del Istmo (CEDI)	1967	?	
Otapan	APSA	1988	1992	
Petapa	APSA	1984	1993	
San Cristobal	Mexican Gulf Sulfur	1954	1957	152
Tehuantepec	Central Minera	1959	1961	42
Texistepec	Cia. Explor, del Istmo (CEDI)	1971	1993	
Poland				
Grzybow	Ciech - Siarkopol	1967	?	

Table 4.10b. Frasch mines – non-U.S. production in 1000 tonnes

Sulfur Domes	Operator	Opened	Closed	Production
Jeziorko (Tarnobrzeg)	Ciech - Siarkopol	1967	2001	75,000
Lubaczow?	Ciech - Siarkopol	1977	?	
Basznia	Ciech - Siarkopol (Sulfur Quest)	1974	1992	
Osiek-Baranow	Ciech - Siarkopol	1993	operating	
Skopanie?	Ciech - Siarkopol	1982	?	
Turkmenistan				
Gaurdark		1971	2000	
Ukraine				
Nimirov		1981	1994?	
Rozdol		197?	1994?	
Yavorov		1974	2000	

(*) Off-shore mine

Chapter 5

RECOVERED SULFUR — ALBERTA

Canada, especially Alberta, has emerged as a global sulfur powerhouse (see Figure 5.1). During the 20th century, 1.5 billion tonnes of elemental sulfur were produced around the world. Half of this amount came from the two largest producing nations, the U.S. and Canada. However, when Canada started producing, the century was already half over. For the past few decades, Canada has been the #1 global sulfur-trading country and maintains that position today.

Canada holds this distinction, though no Frasch or other native sulfur mines have ever operated in this country. The only native sulfur deposits that attracted any serious interest are found in Northern Alberta, near Fort Vermillion (100 miles north of Edmonton). Three times, this sulfur field was evaluated: 1950 - Fortune Oil; 1952 - Sunbeam Sulfur Mines (joint venture between Fortune and Domtar); 1967 - fifty sulfur prospecting permits were issued for the area around Fort Vermilion. However, nothing further came of these exploratory drillings. Where, then, does the Canadian sulfur come from? The source of the yellow element is hydrogen sulfide found in sour natural gas.

There has been a love-hate relationship between sulfur and the oil and gas industry for a long time. The origins can be traced back to the 1880's and the sour oil fields of London, ON, and Lima, OH, which were contaminated with minute but annoying levels of sulfur. High concentrations of hydrogen sulfide were a hidden danger for those exploring for oil. Even the early sulfur mining endeavors at Calcasieu had uncovered deposits of the deadly gas. As wildcatters searched for oil, they, at times, came across sulfur as a consolation prize. Most of the sulfur domes in the first half of the 20th century had been discovered in search for oil, not sulfur. As energy consumption rose, more sour oil and gas fields had to be developed, and sulfur limits were regulated lower. The combination forced the two industries to become inseparable, much to the chagrin of the energy companies.

Pure elemental sulfur is not directly found in natural gas; it is there only in its reduced form hydrogen sulfide (H_2S). These "sour" sources must be treated to remove the corrosive and poisonous gas. The hydrogen sulfide is often trapped using the Girbotol process (amine system). Diethanolamine, monoethanolamine and sulfinol are the common agents to capture the hydrogen sulfide from the gas. The amine then releases a concentrated hydrogen sulfide stream into a multi-stage Claus process, where the hydrogen sulfide is partially oxidized into elemental sul-

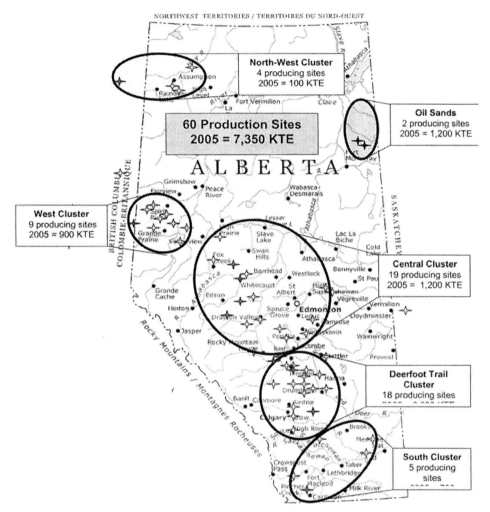

Figure 5.1. Sulfur producing sites in Western Canada. Original map data provided by The Atlas of Canada http://atlas.gc.ca. Copyright 2006. Produced under license from Her Majesty the Queen in Right of Canada with permission of Natural Resources Canada.

fur. The technology was based on the Claus process used by the Leblanc alkali plants of the United Alkali Company back in the late 19th century.

Today, 90% of the supply of elemental sulfur is recovered, which by its diffuse nature globalized the industry. This diversification of sulfur supply started during the 1950's, as oil refineries and natural gas plants began removing sulfur around the world. By 2003, there were more than 700 oil refineries in the world, located in 118 countries; in half of these countries recovered sulfur production took place.

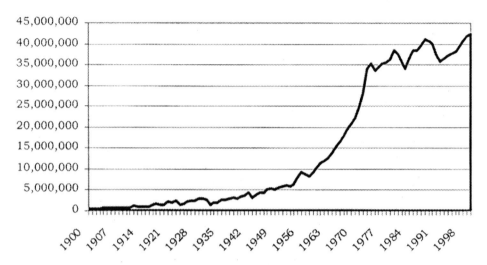

Figure 5.2. Elemental sulfur — World production from 1900 to 2000 (tonnes).

Since the beginnings of the Frasch industry, elemental sulfur production (see Appendix III) grew exponentially until 1974. A notable shift in the growth rate took place at this time (Figure 5.2). Since then growth has continued but at a much slower rate. Two factors contributed to this paradigm shift: the first was the demise of the Frasch industry; the second was the slowing of the explosive growth of recovered sulfur. In 2004, world elemental sulfur production was at a record level of 44 million tonnes, with a value greater than one billion dollars. The largest producing regions are Alberta, Texas/Louisiana, the Persian Gulf and the northern Caspian Sea.

5.1 SOUR, MORE SOUR, & SOUREST

In October 1957, the first sulfur was recovered from oil in Canada. Laurentide Chemical and Sulfur Limited (becoming Sulconam in 1980; now owned by Marsulex) built a recovery plant (100 tonnes per day) for $2.5 million in Montreal East, which was fed by five local refineries. The source of the crude was offshore and, therefore, the sulfur production from this site was not included in official government statistics on sulfur. In the first full year of operation, Laurentide Chemical produced 21,000 tonnes of elemental sulfur. The second such sulfur recovery plant was built by Irving Oil, in Saint John, NB, in 1960. While smaller facilities were later opened (and closed), these two sites have dominated Canadian sulfur production from oil refining. Overall, sulfur production from oil refineries in Canada has been relatively small.

Sour gas wells in Canada are generally found in the Foothills region of Alberta and British Columbia, with the highest hydrogen sulfide content located in Southern Alberta. Although quantities vary significantly from site to site, the average sulfur concentration was ~5 tonnes per million cubic feet of natural gas.

The first sour gas well in Canada was opened in Turner Valley, AB, south of Calgary. The field had been discovered in 1914. In 1924, hydrogen sulfide scrubbing was started at Royalite #4. The recovered hydrogen sulfide, though, was not utilized but was flared or vented. Not until after World War II was there any attempts to recover the sulfur. In the 1949 issue of the **Canadian Minerals Yearbook** (*Canada Department of Mines and Technical Surveys*), it is reported that *Efforts are being made to develop a process to recover elemental sulfur from this* (sour natural) *gas*. The market timing was perfect as a global sulfur shortage loomed on the horizon, as Frasch production could not keep up with escalating demand. Royalite Oil (purchased by BA in 1962) built the second such sulfur recovery plant (30 tonnes per day [tepd]) in June 1952 at Turner Valley.

The first recovered sulfur from natural gas plant in Canada had opened the previous year, in May 1951, at the Jumping Pound, AB, site of Shell, west of Calgary. In its first year of operation, this novel operation in Canada produced just 15 tonnes of sulfur. The following year, recovered sulfur shipments from sour gas from Shell and Royalite amounted to 8,100 tonnes. The initial capacity of Jumping Pound was 15 tepd, which then was steadily expanded: 30 tepd in 1952; 50 tepd in January 1955; 80 tepd in 1956; 240 tepd in 1968. An early consumer of the Shell sulfur was Inland Chemical in Fort Saskatchewan, AB. They opened a sulfuric acid plant in 1955 ($1 million) to supply Sherritt-Gordon; ammonium sulfate was produced as a by-product of their nickel-sulfide ore treatment.

Growth continued through the remainder of the decade (see Table 5.1). Among the new larger sulfur plants was the Pincher Creek facility of Gulf Oil (BA), which supplied liquid sulfur in tank cars to the ammonium sulfate plant of Northwest Nitro Chemicals, in Medicine Hat, AB. The Pincher Creek facility, opening on January 31, 1957, was the largest plant of the day with a sulfur capacity of 225 tepd. This plant was expanded to 690 tepd in 1957 and at the end of the decade still had the largest sulfur recovery capacity in Canada. A few months later, in November 1957, the first B.C. plant was opened in Taylor (300 tepd) by Jefferson Lake Sulfur and Pacific Petroleum. Jefferson Lake Sulfur also formed a joint venture with Mobil Oil (and others), called Petrogas Processing, which opened a recovery plant in Balzac, east of Calgary, in 1961. The only recovered sulfur from natural gas in Canada outside of Alberta and British Columbia opened in 1958 at Steelman, SA (6 tepd; closed in 1980).

Table 5.1. Sour gas plants – Canada (1951 to 1960)

Company	Site	Start-up	H$_2$S, %	Capacity in 1960, tepd
Shell	Jumping Pound	1951	3	100
Royalite	Turner Valley	1952	4	30
Imperial Oil	Redwater	1956	3	9
B.A. Oil	Pincher Creek	1957	10	685
Jefferson Lake Sulfur	Taylor Flats	1957	3	299
Steelman Gas	Steelman	1958	1	6
Texas Gulf Sulfur	Okotoks	1959	35	376
B.A. Oil	Nevis	1959	5	77
Standard Oil	Nevis	1959	6	118
Canadian Oil	Innisfail	1960	14	100

The first sour well with over 10% H$_2$S in Canada was Okotoks (370 tepd) at 35%, which opened on June 3, 1959. By the end of the year, over 50,000 tonnes of sulfur had been produced from this site, south of Calgary. The sulfur was controlled by Texas Gulf Sulfur; other partners at Okotoks included Shell and Devon Palmers Oil. Later, in 1986, the Mazeppa gas plant was opened near Okotoks, by Occidental Petroleum (H$_2$S = 25%). Another of the "sourer" plants was at Crossfield, north of Calgary, by Pan American Petroleum (subsidiary of Standard Oil), opening in 1968 (H$_2$S = 32%). In 1972, Elf Aquitaine (now Total) began operations at Ram River (H$_2$S = 17%), which was the largest sour-gas recovered sulfur plant in the world (from the large Ricinus field). This site retained this position through the decade.

In 1987, Shell discovered the massive Caroline sour gas field (H$_2$S = 25%). Construction of the sulfur recovery plant began in October 1990, and the plant opened in March 1993, just as the sulfur market was falling. Caroline is the largest sulfur recovery plant in Canada, producing 1.75 million tonnes per year, which is three times larger than the second largest producer. The liquid sulfur is sent by a forty-two kilometer underground pipeline ($50 million) to the sulfur prilling (46 Sandvik Rotoformers) and handling plant ($76 million) at Shantz. In May 2005, another major sour gas field was opened by Shell at Tay River, AB, near Rocky Mountain House (H$_2$S = 35%). The sour gas is being processed at Ram River and Strachan.

The "sourest" gas plant in Canada was opened at Harmattan, near Shantz, which had a hydrogen sulfide content of 52%, by Canadian Superior Oil Limited (purchased by Mobil Oil in 1984) in 1966. The site, now owned by Solex, is still operating today. While Harmattan is the sourest natural gas processing plant, even more sour wells have been discovered. In the U.S., a very sour well (H_2S = 78%) was discovered at Black Creek, near Biloxi, Mississippi. Philips Petroleum and Pan American Petroleum investigated the production of sulfur from this site, but decided not to proceed with commercialization after the price of sulfur had fallen in 1969.

The honor of the "sourest" well belongs to Alberta. Shell had investigated an 87% hydrogen sulfide content sour gas well in Bearberry, AB (Panther River), near Caroline, in the late 1950's and early 1960's. A well was drilled in the area by Shell and Canadian Superior Oil in 1969. To call this a sour gas well is a misnomer; it is a hydrogen sulfide well sweetened by natural gas. The site was revisited twenty years later by Shell (along with Mobil Oil, Norcen Energy Resources, and PanCanada Petroleum). The highly corrosive gas requires special metallurgy. A $65-million demonstration plant (200 tepd) was opened by Shell to process the gas in September 1990. Although deemed a technical success, the plant was closed in December 1992 because of economic factors. The sulfur reserves at Bearberry were estimated at one hundred million tonnes. Bearberry is the bitter champion of recovered sulfur sites in the world!

5.2 ALBERTA & THE GLOBAL SULFUR MARKET

With the opening of the first plants, it was already apparent that Alberta would soon be a major sulfur producer. From the meager beginnings in the early 1950's, recovered sulfur production in Canada escalated at an incredible rate, being driven by the demand for natural gas and the "sourness" of Alberta wells. By the end of 1957, total annual recovered sulfur capacity in Canada was still less than 100,000 tonnes. The boom in Canadian natural gas, and associated sulfur, was just about to kick in as pipelines opened up gas exports to the United States. From their yellow crystal balls, Canadian sulfur analysts projected astronomic growth in sulfur production. In the 1955 report of the **Canadian Mineral Yearbook**, T.H. Janes projected that Canadian sulfur production would reach 1.2 million tonnes in 1960, fifty times present production. In 1962, production reached one million tonnes of sulfur from sour gas in Canada, an increase of more than a magnitude in just five years! Janes' daring projection of 1.2 million tonnes was surpassed in 1964. The jump was led by natural gas shipments to the U.S., which had been approved by the newly formed National Energy Board in August 1960. Pipeline construction began immediately. The Alberta Oil and Gas Conservation Board warned about sulfur for the first time in 1963. In their review of 1962, they reported that the sulfur reserves in Alberta gas fields and oil sands, without an "apparent market," were 800 million tonnes.

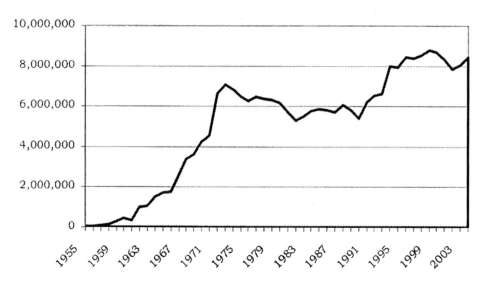

Figure 5.3. Elemental sulfur - Canada production from 1955 to 2005 (tonnes).

Canada reached milestone after milestone as its natural gas production rose, placing this country at the forefront of the global sulfur marketplace:

1959	production surpassed Italy
1962	production surpassed one million tonnes
1964	became the world's largest producer of recovered sulfur, surpassing France (a position that it held until the mid-1980's)
1967	production surpassed 2.5 million tonnes
1968	became the largest exporter of sulfur in the world
1972	production surpassed Frasch sulfur production in the U.S.
1972	production surpassed five million tonnes.

Between 1951 and 1967, more than ten million tonnes of elemental sulfur had been recovered in Canada. By the end of the decade, twenty-five sour gas plants were now operating, mainly in Alberta.

Another bold prediction was made in 1971: Canadian sulfur production would reach ten million tonnes by 1980. For once, the projection was a bit off. Elemental sulfur production in Canada had peaked, and has never surpassed nine million tonnes. By 1980, there were fifty sour-gas sulfur plants operating in Canada, of which forty-six were in Alberta. Sulfur production in this year was 6.1 million tonnes, up from 4.6 million at the beginning of the decade. No new major sour gas finds in Alberta took place until the discovery of the major field in Caroline. For the 1980's, sulfur production was flat, just under six million tonnes per year. Not until 1994 was the production record of 1973 (seven million tonnes) surpassed. Since, Canadian elemental sulfur production has been around eight million tonnes per year. A record of 8.8 million tonnes was produced in 1999 (see Figure 5.3). Canada has been the second largest producer of sulfur, just behind the U.S., for a number of years and remains the largest exporter of sulfur in the world.

5.3 SULFUR BLOCKS & BROKEN SULFUR

Early on a problem developed with the Canadian boom in sulfur production: shipments were not keeping up with production. There were only two choices: cut back on gas production or store the sulfur. Excess sulfur was poured into huge blocks for storage on a scale that had not been seen before. An exceptional feature of sulfur was that it melted at relatively low temperatures (just above the boiling point of water) and was not dissolved by water. Thus, it could be poured as liquid into forms, solidified, and stored outside like giant blocks of yellow ice. The sulfur did not have to be packaged in any way, or protected from the elements. In many respects, it is the easiest chemical in the world to store.

While sulfur may have been the easiest chemical to store, it was not completely problem-free. There are economic issues associated with blocking. The actual blocking can cost several dollars per tonne, and remelting can be in the same range. There are also maintenance costs associated with sulfuric acid run off, caused by bacterial action on the sulfur. Aging blocks are more prone to this problem and to erosion. By their nature blocks are fractured and are susceptible to weathering. A one-million-tonne block may lose a few tonnes of sulfur per year. Environmental costs have been estimated to be $10 million for a three million tonne block. Even after the block has been removed, the base pad remains. In Alberta, the difficult-to-recover sulfur in base pads is estimated to be one million tonnes. Contaminated by the underlying earth and stones, the pads are expensive to recover. An early recovery of a base pad was by Canterra at Ram River in 1987, using a cold flotation process. To recover the 238,000 tonnes of contaminated base pad sulfur took a capital investment of $4.3 million.

Sulfur blocking was not new. In the earliest days of the modern sulfur industry, Union Sulfur was pouring sulfur into wooden forms that were 150 ft. x 250 ft x 65 ft, containing more than 100,000 tonnes of sulfur each; another innovation of Herman Frasch. By the 1930's, the larger blocks were now 1,200 feet long. Wooden supports were still being used, but galvanized steel sheets had also been introduced. The sulfur was then recovered by blasting the friable blocks and shipping out the "broken" sulfur in hopper cars. By 1950, dynamite was still a popular choice, but massive power shovels were also used to gouge out the yellow element from the blocks.

Broken sulfur was replaced by liquid (or liquid into prills) because of environmental and safety issues. An early attempt to melt solid sulfur in blocks was made by Freeport Sulfur at Joliet, IL, in 1964. In 1977, the first melters were uti-

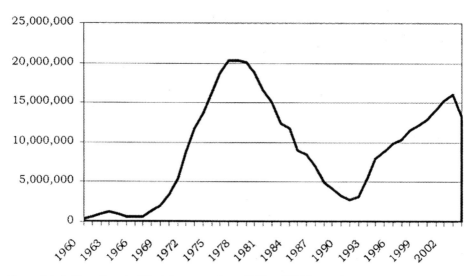

Figure 5.4. Sulfur - Canada. Block inventory from 1960 to 2005 (tonnes).

lized to recover sulfur from blocks in Canada. These devices heat the top and out-side of the block, shaving layers off the yellow ice palisades. Melting capacity in Canada had reached 300,000 tonnes per month three years later.

The inventory problem in Canada first began to manifest itself as early as 1958, when shipments of elemental sulfur were 86,000 tonnes, but production was almost 140,000 tonnes. By 1962, block inventories were over one million tonnes. During the 1970's, sulfur inventories in Canada grew at an alarming rate as pro-duction continued to outpace demand. Some projections had block inventories reaching 50 million tonnes by the end of the decade! The sulfur blocks soared from 600,000 tonnes in 1967 to 20 million tonnes in 1977! Finally, the trend started to level off and the inventories peaked at 21 million tonnes in July 1979, which was more than 50% of the world's annual demand. What appeared as a great burden to the industry, amazingly transformed itself into a gold mine, as sul-fur pricing hit record high levels. This huge inventory quickly dissipated in the volatile world of the sulfur industry, passing below the ten million tonne level in 1985, eventually reaching a low of 2.8 million tonnes in 1991 (see Figure 5.4).

While total block inventories in Alberta have risen since 1991, the trend is misleading. For the past few years, sour gas sulfur blocks have steadily declined. For example, the largest block inventory in Canada (outside of the oil sands) has been at Ram River, which by 2002 was over three million tonnes. Since then, the block inventory at Ram River had declined to 1.7 million tonnes by mid-2006. Between 2001 and 2005, the total sour-gas sulfur inventories have dropped by 3.5 million tonnes. This decline, however, has been offset by the block inventory of

Syncrude from its oil sands operations. In 2005, this inventory alone stood at six million tonnes. In other words, the readily accessible sulfur inventories have steadily declined, but this decline has been outpaced by increases in remote inventories that are more difficult to market (because of logistic costs). This is not just a Canadian phenomenon. The major remote blocks are Syncrude and Tengiz.

5.4 SULFUR FORMING: SLATE & PRILLS

In the earliest days of the industry, liquid sulfur was allowed to solidify into blocks and was then dug out by hand or mechanical shovel, or blown apart by dynamite. The sulfur chunks are friable, forming dust easily. The dust from "broken" sulfur was more than just a nuisance, as it posed a dust-explosion hazard. Broken sulfur shipments largely disappeared after liquid shipments took over after World War II. However, Alberta was too far from ocean ports to justify liquid sulfur movements overseas. While Canada ships over one million tonnes of liquid sulfur by rail car into the U.S. every year, it has never shipped liquid sulfur by vessel. A major project was undertaken to develop such a mode of transportation in the late 1990's. The Sulfur Corporation of Canada began construction of a liquid sulfur terminal at Prince Rupert, BC, in 1999, but the company ran out of cash in July 2002, and was forced to abandon its project.

So far, we have discussed the marketing of sulfur in two states, but there is a third "form:"

I.   Broken sulfur from blocks
II.   Liquid sulfur directly from the process or melted from blocks
III.   Formed sulfur from liquid sulfur

Sulfur forming was invented to allow the transportation of solid sulfur with less risk than broken. The earliest sulfur forming was flake or slate sulfur. Liquid sulfur is poured onto a water-cooled moving belt to produce a sheet or ribbon of sulfur, which falls off the belt and breaks into chips, called slate. In the mid 1950's, Texas Gulf Sulfur was experimenting with slate sulfur. A plant was installed at their Chacahoula mine. One of the earliest sulfur slate-forming plants was installed by Pemex of Mexico in 1953; the unit contained a water-cooled belt 32 inches wide and 184 feet long. The early formed sulfur of Pemex was three to four inches long and ¼ inch thick.

In the 1960's formed sulfur was introduced into Canada. From July 1972, only formed sulfur could be shipped in bulk from Vancouver by an order from the Vancouver port authorities, for safety and environmental reasons. Annual forming (slate) capacity at this time was three million tonnes. In 1974, a large slate plant was installed by Shell at Waterton, AB, and, in 1977, a major Sandvik slate plant was installed at Ram River (Elf Aquitaine). By 1980, slate capacity had been expanded to five million tonnes.

In the early 1970's, another method of sulfur forming was introduced, prilling (i.e., pelletizing). In 1974, the U.S. Bureau of Mines reported (**Minerals Yearbook**):

> The use of slate sulfur for bulk overseas shipments has partially alleviated the problem of dusting, but it has not been entirely effective in this respect. It has been proposed that pelletizing or prilling of elemental sulfur would provide a more satisfactory product for large-scale movement in international trade. Several relatively small plants for the production of pellets or prills have been built to evaluate these products from the standpoint of shipping characteristics, dusting problems and customer acceptance.

Generally, prills produce less dust than slate, and therefore are better for overseas trade. The earliest sulfur prills can be traced back to Agricola in his **De Natura Fossilium**, where he reported that sulfur "seeds" were produced by "distilling sulfur drop by drop through the openings of jars." During the 1930's, a similar form of sulfur was produced by the ORKLA process, where the sulfur was the size of a grain of corn. The modern prilling technologies began to emerge in the late 1950's; SNPA of France had tested pelletizing technology from Scandinavia at Lacq. European companies led the development of the early technologies, which included, Sulpel, Kaltenback, Ciech (Polish), and Perlomatic (French) processes. The first two processes use water, while the later two use air, to cool the sulfur. In 1964, the wet prilling process was reported where liquid sulfur droplets were cooled by a water vortex, producing what is sometimes referred to as "popcorn" sulfur. Another process was the Rotoformer designed by Sandvik. A Canadian development was the "Procor GX" dry granulation process. In this process, sulfur is sprayed in a rotating drum containing sulfur "seeds."

By the later 1970's, prilling technology began to be installed by Canadian sulfur producers:

1976 –   Perlomatic – Balzac by Petrogas
1977 –   Sandvik – Ram River by Aquitaine
1978 –   Procor – Windfall by Texas Gulf Sulfur
1978 –   Procor – Harmattan by Shell
1980 –   Ciech – Strachan by Gulf Canada
1980 –   Fletcher (wet prilling) – Fort Nelson and Taylor, BC, by Westcoast Transmission
1980 –   P.V. Commodity Systems – Strachan by PVC
1980 –   Procor – Fort McMurray by Suncor and Syncrude
1981 –   Fletcher – Pine River by Westcoast Transmission

1981 –   Ciech – Waterton by Shell
1981 –   Ciech – Kaybob by Hudson Bay Oil and Gas
1981 –   Ciech – Ram River by Aquitaine
1984 –   Procor – Crossfield by Amoco

The prilling process quickly replaced slate. By 1981, 60% of the off-shore export shipments from Canada were prilled. The only remaining major slate producer in Canada is Husky Energy. In 2005, total forming capacity was eight million tonnes per year in Canada (see Table 5.2), and greater than 30 million tonnes per year worldwide.

Table 5.2. Forming capacity – Canada

Company (location)	Capacity, tepd
Shell	
Shantz	5,150
Waterton	2,100
Jumping Pond	1,920
Husky	4,600
Duke Energy	
Pine River	3,000
Fort Nelson	1,000
Central Alberta Midstream (SimCanada)	
Kaybob 3	1,050
Kaybob South	300
PetroCanada	770
Nexen	1,575
Keyera	1,000
Suncor	?
Talisman	550
Newsul	400
Tiger	264
Canada	**23,679**

5.5 MARKETING & LOGISTICS

5.5.1 CANSULEX & PRISM

As the number of sour gas plants escalated, so did the number of competitors. Canada had to start selling their sulfur, and competing against the Frasch producers was no easy task. In the early days of the industry, Canadian producers had two serious competitive disadvantages:

• Broken sulfur: while the rest of the world was going liquid, Canadian producers had no liquid network; already at a logistics disadvantage, they could ill-afford the expense of a liquid terminal network on the west coast.

- Logistics: to get the sulfur overseas, the sulfur had to be transported to Vancouver, while the Frasch producers were already near ocean-going ports.

While forming reduced the former problem, the latter was a far greater challenge.

The U.S. had long ago eliminated export competition among themselves by forming Sulexco. The Canadian industry followed a similar path. In 1961, Cansulex was formed by BA and Petrogas to handle sulfur exports. Petrogas, itself, was a consortium of sour gas producers in Alberta, including Jefferson Lake Sulfur, Mobil Oil, and 26 other producers. The other major overseas exporter at the time, Shell, was selling through the International Sulfur Company. A smaller independent Canadian trading firm was Brimstone Exports Limited (formed by Raymond Learsy in 1963; in 1992, Brimstone declared bankruptcy).

By 1980, Cansulex had 23 members, which encompassed 65% of Alberta's sulfur capacity. Shell, Canadian Superior Oil (owned by Mobil Oil), and Amoco were the major non-members. Mobil, itself, withdrew in 1987. From 1978 to 1990, Robert Q. Phillips (~1926 - 2001) was head of Cansulex. During his tenure, sales increased from $66 million to over $200 million. When Phillips retired, he was replaced by David K. Arnott, who had been executive vice-president. Cansulex folded by the end of the year, and Arnott later joined Vancouver Wharfs.

In late 1991, a new sulfur exporting group was formed called Prism Sulfur Corporation, which, at first, controlled 90% of Canadian exports of sulfur. The new consortium officially took over exports on January 1, 1992. The first president of Prism was D. Michael G. Stewart, who was followed by Doug Stoneman and, in 1995, by Walter Litvinchuk. The dissolution of Cansulex and the emergence of Prism may have been undertaken to encourage the major sulfur producers that were not part of the former company to join the new organization. If this was the purpose, then it did not work for long. Membership peaked in 1993 with thirty-three companies, but had fallen to twenty-one in 1995. Of the ten original members, three of the largest, Shell, Amoco and Husky Oil, along with smaller companies, withdrew at the end of 1994. In 1998, Bruce W. Coulbourn became president of Prism and was succeeded in 2001 by F. Terry Draycott, who had been president of Enersul.

5.5.2 ALBERTA TO VANCOUVER

For Cansulex to compete against Sulexco, the Canadian sulfur had to get to port at a similar cost as the Frasch price in the Gulf. Whereas the Frasch companies were already at the Gulf, Albertan sulfur had to be shipped hundred of kilometers to the port of Vancouver. Essential to the success of the Canadian sulfur industry was an efficient logistics system, as first noted in the **Canadian Minerals Yearbook** in 1958 and repeated in later issues. Various options were investigated. Pipelines were well known to the industry, and projects were contemplated to ship liquid or slurried sulfur. In 1960, Pembina Pipeline proposed a 1300-mile, 14-inch pipeline to carry sulfur slurry from Alberta to Chicago. Four years later, a 750-mile pipeline from Alberta to Vancouver was proposed to transport liquid or slurried sulfur. In 1966, Shell, through its subsidiary Commercial Solids Pipeline Company, announced a similar project, a 12-inch liquid sulfur pipeline from Southern Alberta to Vancouver. The plans for the $50 million project were later dropped. Such major pipeline plans were seriously considered, but never implemented.

A more immediate approach was to lower the rail rates. The port of Vancouver is serviced by CP Rail. The rail company was open to lower rates so that traffic would rise. In 1961, the rail rates were slashed: to Chicago, from $20 to $13 per tonne; to Vancouver, from $10 to $7.75 per tonne. The U.S. Frasch producers watched these changes with suspicion. Freeport Sulfur challenged the legality of these cuts, accusing the government of indirect subsidies. A U.S. court upheld the rate tariffs. Later in the decade, a new rail strategy was introduced to cut rates even more by shipping massive single rail movements. In 1969, Shell first used "unit trains" to take elemental sulfur to the port of Vancouver. Rail costs were reduced to $5 per tonne with the unit trains. In cooperation with the railroads, the sulfur producers had managed to cut the freight rate to Vancouver by 50% by the end of the decade.

Vancouver, one thousand kilometers from the major sulfur production sites in Alberta, has always been the major port for sulfur in Canada. The sulfur terminal was built in 1960. Newly formed Cansulex wasted no time in establishing Canadian sulfur onto the global scene. The first bulk shipment of sulfur (10,000 tonnes) was shipped out of Vancouver to Taiwan in October 1961. By the end of 1962, Cansulex was selling Canadian broken sulfur to the Far East, South-East Asia and India. Sulfur exports soared from Canada. In 1965, a special movement of 26,000 tonnes was loaded on board the vessel *Grimland* from almost 400 rail cars. The special nature of this shipment was not just its significant size but that it went to Europe. By 1968, Canada was already the largest exporter of sulfur in the world, controlling about one-third of global trade. Canadian sulfur first went to the U.S. East Coast by vessel out of Vancouver in 1977. During the 1980's, Canada dominated the export market, accounting for an amazing 40% to 45% of global trade. As recovered capacity around the world increased, especially in Russia and the Persian Gulf, and with production remaining relatively stagnant, Canada's market share deteriorated. In 2004, world trade in elemental sulfur was over 25 million tonnes, with Canada exporting 8.3 million tonnes.

With such massive movements, the availability of hopper cars to deliver to the port became an increasing issue. In 1976, Sultran was formed by the sulfur producers to operate the rail fleet, manage the bulk sulfur movements by hopper cars to Vancouver, and to negotiate rail rates. In 1981, Sultran added 680 rail cars to their fleet to expedite the shipments of sulfur to Vancouver and purchased PCT in Port Moody from Cominco. Pacific Coast Terminal (PCT) is one of two sulfur terminals in the Vancouver area; the other is Vancouver Wharves (purchased by BC Rail in 1993). PCT dredged their berth to accommodate Panamax-sized vessels in 1995.

Scheduling of vessels to the arrival of unit trains was nearly impossible. Large inventories of sulfur had to be kept at the port. The formed-sulfur inventory would have to be communal, since most producers were shipping through Cansulex, and later Prism, to the port. In 1970, the Solid Sulfur Train Operating and Exchange Plan established a common inventory for twenty-four sulfur producers at Vancouver. The sulfur was segregated into three piles: slate, rotoform (Shell), and other prill.

However, the Vancouver facilities have often been unable to keep up to the escalating demand for sulfur, forcing exporters to other ports. For example, in 1974 and 1975, 100,000 tonnes of sulfur were shipped out of Churchill, MB (Amoco), Thunder Bay, ON (Shell), and Quebec City, QC (Suncor). In the following year, 150,000 tonnes were shipped out of the latter two ports, but shipments through Churchill had ceased the previous year. In 1977, the non-Vancouver shipments had ended, but two years later, exports out of Thunder Bay restarted (150,000 tonnes in 1980) and small volumes were shipped out of Prince Rupert, BC, which increased to 100,000 tonnes the following year. A Fletcher process sulfur granulator was installed at Prince Rupert by Real International Marketing Limited in 1981. Vancouver, though, remains the Canadian sulfur gateway to the world.

5.6 THE GLOBAL COMPETITION

5.6.1 U.S.

In the early 1950's, the **President's Materials Policy Commission** issued a report called **Resources for Freedom**. This presidential body made the bold projection that sulfur demand in the U.S. would reach 10.1 million tonnes by 1975. Demand in 1950 had been only 5.1 million tonnes. The projection was almost perfect. The 10.1 million level was first surpassed in 1973; demand in 1975 was 10.8 million tonnes! The Commission also issued a warning (**Minerals Yearbook** 1952):

> After reviewing the supply situation, the President's Materials
> Policy Commission published estimates of reserves and con-

cluded that Frasch reserves are not large enough to supply the growing requirements for many years and by 1975 other sulfur minerals would become increasing important sources of supply.

While the Commission was correct about the Frasch supply, their conclusion of "other sulfur minerals" was off the mark. Frasch sulfur would be largely replaced, not by other minerals, but other sources. The new source was already there. As early as 1937, the potential of recovered sulfur was reported by Gustav Egloff (1886 - 1955) of the Universal Oil Products Company (Haynes, **The Stone That Burns**, p. 261):

> The hydrogen sulfide available from natural gas, cracking-still gas and refinery gas, along with the sulfur in crude oil, amounts to more than enough to produce the entire sulfuric acid needs of the nation.

Before Canada had produced one tonne of recovered sulfur, the U.S. was already the largest producer in the world, reaching 140,000 tonnes in the U.S. in 1950 and surpassing 500,000 tonnes, seven years later. Most recovered sulfur in the U.S. came from oil refineries. By the end of the 1950's, recovered sulfur producers were beginning to impact traditional Frasch sulfur markets. The one-millionth tonne milestone for recovered sulfur from oil and natural gas was reached in 1964. By 1970, there were almost one hundred producing sites across twenty-three states; five companies accounted for over 40% of the recovered sulfur production in the U.S. (see Table 5.3).

The Texas/Louisiana region remains a major sulfur producer, no longer Frasch, but recovered. By 1977, over one million tonnes of sulfur was being recovered in Texas from oil refineries. Total recovered sulfur from only oil refineries had surpassed the one-million tonne mark in 1972 and two million tonnes in 1979. By 2002, over 40% of the global recovered sulfur capacity from oil refineries was found in the U.S.

The first major sour gas plant in the U.S. was opened by Southern Acid and Sulfur Company (later becoming part of Olin in 1949) at McKamie, AR (120 tepd). They opened another plant in Magnolia, AR, two years later. Texas Gulf Sulfur was also an early pioneer in this field. Their investigations started in 1941, and they opened a pilot plant for recovering sulfur from sour gas also at McKamie. In 1949, Texas Gulf Sulfur built a recovered sulfur plant (400 tepd) at Worland, WY, for Union Oil (Pure Oil), which was the largest recovered sulfur plant from sour gas in the world. Worland closed in 1967 after producing over one million tonnes of recovered sulfur. By 1950, there were six facilities recovering sulfur from sour gas in the U.S.:

- Texas Gulf Sulfur: Worland, WY
- Freeport Sulfur: Eagle Point, NJ
- Olin: McKamie, AR; Magnolia, AR
- Hancock Chemical: Los Angeles, CA
- Stanolind: Elk Basin, WY

Recovered sulfur from sour gas surpassed the one-million-tonne milestone in 1973, hitting two in 1983. Although there was significant growth in natural gas discoveries in the U.S., most were sweet wells. During the early 1970's, sour gas plants were built in Florida, Mississippi and Alabama. Sulfur production in Wyoming surpassed one million tonnes for the first time in 1987.

The recovered industry in the U.S. hit a new milestone in 1998, when it surpassed the record Frasch production (1974 = 7.9 million tonnes). Another passing of the baton took place in 1999, when Freeport Sulfur was surpassed by Exxon-Mobil as having the largest sulfur capacity in the U.S. In 2004, nine million tonnes of recovered sulfur was produced, with almost one-half originating from Texas and Louisiana. The U.S. remains the largest producer of elemental sulfur, even with the demise of the Frasch industry, as recovered sulfur dominates. The U.S. is also still the largest consumer of sulfur, outpacing its supply.

Table 5.3a. Recovered sulfur – U.S. top five producers

1970	1975	1980	1985
Cities Service Oil	Exxon	Atl. Richfield	Chevron
Getty Oil	Getty Oil	Chevron	Exxon
Pan American Petro.	Mobil	Exxon	Shell
Shell	Shell	Shell	Standard Oil
Stauffer Chemical	Standard Oil	Standard Oil	Texaco
1990	**1995**	**2000**	**2005**
Chevron	Exxon	ExxonMobil	ExxonMobil
Exxon	Standard Oil	BP	BP
Shell		Chevron	Chevron/Texaco
Standard Oil	Mobil	Motiva	Motiva Enterprises
Star Enterprise	Shell		ConocoPhillips

Source: **U.S.G.S. Minerals Yearbook**

Table 5.3b. Recovered sulfur – U.S. 2005 producers

Company	Sites #	Capacity, 1000 tepy
Exxon/Mobil	13	2,005
BP	17	1,320
Chevron/Texaco	8	1,185
Motiva Enterprises	5	800
ConocoPhillips	10	695
Shell	10	640
Marathon Ashland Petroleum	6	600
Pursue Energy	1	580
Valero Energy	10	555
CITGO Petroleum	3	365
Others	95	3,695
Total	**178**	**12,440**

Source: *Chemical Marketing Reporter*, **268** (10), p. 38, (2005)

5.6.2 WESTERN EUROPE

In Western Europe, France was the first major producer. Recovery of sulfur from oil began in 1954 at the Berre-l'tang refinery. On April 21, 1957, the natural gas fields in Lacq (H_2S = 15%), 80 km from Bayonne, were opened by SNPA (Societe Nationale des Petrole d'Aquitaine; now part of Total). Total sulfur production in this first year was only 28 tonnes. French sulfur had surpassed the one million tonne mark in 1960, surpassing the U.S. as the largest producer of recovered sulfur in the world. In 1963, French sulfur (7,000 tonnes) appeared in the U.S., and the following year, liquid French sulfur was sent by vessel from Rouen to Immingham, in the UK, using the *President Andre Blanchard* (named after the head of the sour gas project). French production peaked in 1978 at over two million tonnes. Since then it has declined and now sits at one million tonnes per year (see Table 5.4).

By the middle 1950's, Germany (East and West) was producing more than 150,000 tonnes of recovered sulfur per year. In 1979, recovered sulfur from West Germany surpassed the one million tonne level. Germany had become the second largest sulfur producer in Western Europe, after France. The major region was the sour gas fields of Lower Saxony.

Liquid sulfur terminals were opened across Western Europe: Rotterdam (Sulexco, SNPA; Siarkopol), Botlek, Antwerp (APSA; Sulexco), Rouen (SNPA),

Immingham (SNPA, APSA), Teeside, Runcorn, and Workington in the U.K., and Seaport Brake (NEAG) in Germany.

Table 5.4. Recovered sulfur – Western Europe production (1000 tonnes)

Year	France	Germany
1960	778	193
1965	1,497	208
1970	1,708	280
1975	1,962	840
1980	2,216	1,384
1985	1,723	1,779
1990	1,050	1,175
1995	1,170	1,000
2000	1,150	1,753

(*) German data includes some metallurgical sulfuric acid; official data listed above is reported to understate significantly total German sulfur production)
Source: **U.S.G.S., Minerals Yearbook**

5.6.3 PERSIAN GULF (SEE TABLE 5.5)

5.6.3.1 Iran

The **U.S. Minerals Yearbook** first mentioned, in terms of sulfur, the Persian Gulf area in 1945, when it reported that there were various small native sulfur mines in the region, and a summary of native mines in Iran were reported in the **Minerals Yearbook** of 1947. Just after World War II, sulfur was being recovered from oil refineries at Abadan, but by the late 1950's, only 15,000 to 20,000 tonnes of recovered sulfur was being produced each year. Iran was the dominant recovered sulfur producer in the region throughout the 1970's with the opening of major recovered sulfur sites:

1969	National Iranian Oil and American Oil at Kharg Island ($H_2S = 12\%$);
1970	Shahpur Chemical (j.v. Allied Chemical and National Iranian Oil) at Shahpur (Bandar Khomeini) ($H_2S = 25\%$); the sulfur-removing technology was licensed from SNPA of France;
1986	Khangiran, near Sarakha;
1991	Hasheminejad.

5.6.3.2 Iraq

Native sulfur was being mined by the Iraq Sulfur Company at Kadhimiyah. At this site, near Baghdad, they used the infamous *Calcarone* kilns. During the early 1950's, Texas Gulf Sulfur was in discussions with the government to develop the sulfur industry of Iraq, but negotiations broke off in 1955. Other American and British firms were also studying the recovery of sulfur. In 1958, Ralph M. Parsons & Company was commissioned to build a 300 tepd sulfur recovery plant at the sour gas plant at Kirkuk. Iraq National Oil opened a new recovered sulfur plant at this field in 1984 and another facility opened at the petroleum refinery at Beiji. Iraq began producing major sulfur volumes, with the largest source remaining the Frasch mine at Mishraq.

5.6.3.3 Saudi Arabia

Occidental Petroleum built the first recovery plant in Saudi Arabia at Damman in 1968. Major sulfur production did not begin until 1980. The introduction of Saudi Sulfur to world markets began in March 1982 after the opening of a prilling plant at Jubail. The first major shipment took place in April, with 23,000 tonnes of prill going to Tunisia. Before this time, sulfur had gone only to block. Inventories had quickly accumulated to over one million tonnes. By 1984, Saudi Arabia was the third largest exporter of sulfur in the world, ahead of the United States and Mexico, and had moved all the previously blocked sulfur. Saudi Sulfur (owned by Devco) installed a Fletcher process prilling plant at Berri in 1984, after the Jubail prilling facility was destroyed by a fire in October. During the 1980's, the major sources of sulfur were the sour gas plants at Uthmaniyah, Shedgum, and Berri, and the oil refinery at Ras Tanura. Exports were handled by the Saudi Sulfur Company (SASULCO) until 1989 when they were transferred to Samarec, and four years later, the rights were bestowed upon Saudi Aramco. In the early 1990's, Saudi Arabia was still the third largest sulfur exporter in the world, after Canada and Poland, and by 2005, prilling capacity had expanded to 3.7 million tonnes per year.

5.6.3.4 Other

Kuwait   Al-Ahmadi, Mina Abdulla, Shuaibai;
Abu Dhabi   Das Island (National Oil Company);
Qatar   Umm Sa'id.

Table 5.5. Recovered sulfur – Persian Gulf production in 1000 tonnes

	1960	1965	1970	1975	1980	1985	1990	1995	2000
Iran	20	20	405	467	150	150	635	840	963
Iraq				108	40	70	380		

Table 5.5. Recovered sulfur – Persian Gulf production in 1000 tonnes

	1960	1965	1970	1975	1980	1985	1990	1995	2000
SA*			5	3	460	1,100	1,435	2,400	2,101
Kuwait			47	54	120	238	300	559	512
Bahrain				24	33	36	5		
AD**						105	90	257	1,120
Oman						14	30		
Qatar						37	52		

* Saudi Arabia, ** Abu Dhabi
Source: **U.S.G.S. Minerals Yearbook**

5.6.4 CASPIAN SEA

Total recovered sulfur production in the U.S.S.R. grew rapidly after 1970:

1970	0.4 million tonnes
1975	1.0 million tonnes
1980	1.7 million tonnes
1985	2.0 million tonnes

Since the latter 1980's, Russia has been the third largest elemental sulfur producer, after the U.S. and Canada. The major sour gas fields are located in the vicinity of the Caspian Sea.

In 1974, a sulfur recovery plant was opened at the Orenburg sour gas field. By 1978, over one million tonnes per year of recovered sulfur were being produced from the site. Gas reserves started to decline. However, the discovery of sour gas in neighboring Karachaganak, Kazakhstan, added new reserves for the facility to process.

The largest point source of recovered sulfur in the world is the Astrakhan Gas Processing Plant, owned by Gazprom, in Russia. Construction of the gas plant began in 1981. The sulfur recovery technology was supplied by Technip of France, who had earlier provided technology for the Orenburg complex. The first phase of the Astrakhan sour gas ($H_2S = 24\%$) project at Aksaraysk was completed in late 1986, when the first sulfur was produced. After five years, the plant was still being commissioned, with only three of four production lines fully operational. Sulfur production surpassed the one-million tonne level in 1992 and continued to increase dramatically afterwards: 1995 = 1.7 million tonnes; 2001 = 3.8 million tonnes; 2002 = 5.0 million tonnes. The site contains a 2000 tonne-per-day pelletizing plant from Devco (installed in 1999). Formed sulfur from the plant is shipped out of the port of Novorossiysk, on the Black Sea, or Mariupol, on the Azov Sea, under the control of Fedcom. Exports in 2004 were 4.3 million tonnes. In 2000, a liquid sulfur terminal had been under study for this port.

Nearby to Astrakhan in the North Caspian Basin are the Tengiz sour gas fields in neighboring Kazakhstan. Development of the Tengiz fields ($H_2S = 25\%$; now 12.5%) of Western Kazakhstan began in the mid-1980's. A French consortium, led by the engineering firm Lurgi, installed the technology. In June 1985, there was a massive blowout in the field. The high hydrogen sulfide levels made controlling the fire difficult. The fire was still burning at year end and was not put out until March 1986. Sulfur production had begun in the spring of 1991. In April 1993, a joint venture between the government of Kazakhstan's oil and gas company Tengizmuniagaz (Kazakhoil) and Chevron was formed, Tengizchevroil (TCO). By 1996, sulfur production was 500,000 tonnes per year, surpassing one million tonnes by 2000. Some slate sulfur was shipped into Western China, but the bulk of production went to block, with inventories reaching 6 million tonnes by 2002. In 2003, new prilling facilities (Caspian Sulfur Corporation) were opened (capacity = 800 ktpy), replacing the earlier slate unit. ICEC was the selling agent for the prills. Shipments increased, going to Israel, Tunisia and Spain, but still much product was put into inventory, reaching 8 million tonnes in 2005. The major exit port is Illychevsk. In 2005, the ownership of TCO was Chevron (50%), ExxonMobil (25%), Kazakhoil (20%) and Lukarco [BP/Lukoil j.v.] (5%).

5.6.5 JAPAN

Through the late 19th century, Japan was the second largest sulfur producer in the world. A notorious mine, located near Kushiro (opened 1885), used prisoners to mine the sulfur. Within six months, forty-two of the prison workers had died. The conditions were so harsh that the government closed the mine in 1888. One of the most spectacular examples of the early Japanese industry was at the geyser-like craters in Bungo province. Pipes were placed in the ground to channel the sulfur into reservoirs; this "Bungo sulfur" was then sold. Native mines were also found at Matsuo, which was opened in 1914 and Horobets. Japanese production had surpassed 100,000 tonnes for the first time in 1916.

The largest liquid sulfur fleet, although the vessels are small by trans-Atlantic standards, belongs to Japan (see Table 5.6). Increasing recovered sulfur production had led to Japan being in an excess supply situation by the mid-1970's. The first ship was commissioned in 1977 and a few more in the 1980's, but the majority were launched during the 1990's. The newest of these vessels is the *Hestiana* launched in 2005. Japanese sulfur suppliers (including Mitsubishi, Itochu and Mitsui) established a liquid network in China, Philippines and India, during the 1990's. By 2001, recovered liquid sulfur from Japan to China was over 700,000 tonnes. The initial system was crude, whereby liquid sulfur was poured into pits to harden. The solid was then broken up to be shipped to clients by rail in bags. The major users of sulfur near the sulfur terminals have recently shifted to liquid tank truck deliveries. In the Philippines, the *Lucretia* is used to make deliveries to the liquid terminal of Philphos at Isabel, built in 1997. From the two 8,000 tonne storage tanks, a one-kilometer pipeline takes the product to the phosphate plant. In India, liquid sulfur was being supplied to Coromandel Fertilizers using the *Sulfur Global*. Many of the Japanese vessels have been manufactured by Samho Shipping of Korea, which have produced thirteen 3,000 DWT liquid sulfur carriers.

Table 5.6. Liquid sulfur vessels

Vessel name	Chartering company	Size, DWT	Sulfur, tonnes	Year
Sulfur Queen	Texas Gulf Sulfur	15,260		1961
Louisiana Sulfur	Freeport Sulfur	16,000		1961
Etude	APSA (Mexico)	15,993		1961
Pochteca	APSA (Mexico)	10,500		1962
Naess Texas	Sulexco	25,300		1964
Naess Louisiana	Sulexco	25,300	23,700	1964
President Andre Blanchard	SNPA (France)	11,000		1964
Marine Texan	Texas Gulf Sulfur	23,800		1964
Louisiana Brimstone	Freeport Sulfur	24,000		1965
Otapan	APSA (Mexico)	23,428	20,000	1965
Marine Floridian	Texas Gulf Sulfur	25,236		1967
Marine Duval	Duval Sulfur	24,734	24,000	1969
Stella Duval	Duval Sulfur	3,255		1971
M.S. Norvest	Siarkopol (Poland)	5,500		1971
?	SNPA (France)	?		1972
Professor K. Bogdanowicz	Siarkopol (Poland)	9,000		1974
Zaglebie Siarkowe	MG Chemiehandel	9,783		1976
Nam Hae Sulfur	(Japan)	3,000		1977
Texistepec (ex. N.R. Crump)	APSA (Mexico)	28,939		1979
Sulfur Pioneer	(Japan)	1,268		1982
Sulfur Frontier	(Japan)	1,605		1984
Kokamaru	(Japan)	937		1986
Teoatl	APSA (Mexico)	21,364	20,000	1989
Sulfur Mercator	(Japan)	1,225		1990
Saehan Sulfur	(Japan)	1,605		1990
Namhae Pioneer	(Japan)	3,000		1991
Nordic Louisiana	Freeport Sulfur	26,500		1992
Benno Schmidt	Freeport Sulfur	8,188	7,500	1992

Table 5.6. Liquid sulfur vessels

Vessel name	Chartering company	Size, DWT	Sulfur, tonnes	Year
Janana	ADNATCO (Dubai)	9,365		1993
Sulfur Glory	(Japan)	3,000		1993
Sulfur Enterprise	Freeport Sulfur	27,240	25,300	1994
Tenshinmaru	(Japan)	1,600		1994
Sulfur Espoir	(Japan)	4,000		1995
Sulfur Tripper	(Japan)	1,156		1995
Kaliope	MG Chemiehandel (Germany)	15,340		1995
Sulfur Global	Mitsui (Japan)	11,872	6,270	1996
Penelope	MG Chemiehandel (Germany)	15,239		1996
Lucretia	Mitsubishi (japan)	10,006	9,300	1997
Kohshinmaru	(Japan)	4,287		1998
Sulfur Spirit	(Japan)	5,500		1999
Mitrope	MG Chemiehandel (Germany)	15,718		1999
Aurora	PCS	24,668	20,000	2000
Morning Sea	(Japan)	3,000		2002
Hestiana	Mitsubishi (Japan)	3,600		2005

5.7 SULFUR PRICING: THE PENNY SYNDROME

An amazing, disturbing if you are a supplier, feature of sulfur is the resiliency of its price, remaining within a relatively narrow range over the 19th, 20th, and now the beginning of the 21st century! The penny-a-pound barrier acts like a magnet, some may say a "yellow hole," which continually draws sulfur pricing to this level. For 200 years, during "normal" times, pricing varied from $18 and $25 per tonne, f.o.b. ex. production site (all pricing is in U.S. funds, unless otherwise stated). In periods of strong markets, the price briefly went up to $30 per tonne, and when markets were depressed down to $15 per tonne. While there have been short periods outside of this range, they were rare. The only extended period of abnormal pricing began in 1975 and continued until 1995 (see Figure 5.5).

As the sulfur industry was coming into its own in the first few decades of the 19th century, sulfur pricing from Sicily was equivalent to about $20 per tonne. In 1831, speculators had pushed the price up to an unprecedented $70. After a brief period of volatility, pricing returned to normal. Stability did not last long as the TAC scandal had forced pricing briefly higher again. Following the Sulfur War, sulfur prices had a long-period of relative stability back to the $20 per tonne

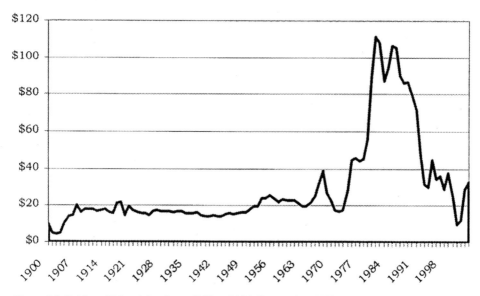

Figure 5.5. Sulfur – U.S. pricing from 1900 to 2004 (f.o.b. mine, US$ per ton)

range. Besides the volatile 1830's, the price of sulfur seldom varied during the entire 19th century.

A famous incident in the archives of sulfur pricing took place in 1898. Tensions were rising in the days after the sinking of the Maine in Havana on February 15th. As Spain and the U.S. came closer to war, there was speculation that sulfur shipments would become regulated by the government. A Sicilian freighter lay in New York City loaded with sulfur. The original f.o.b. price was only $16 per tonne. However, the war rumors soon spread and a bidding war took place over the cargo. The sulfur was eventually purchased by Herman Ridder (1851 - 1915), owner of the German-language newspaper *Staats-Zeitung* (his company merged to form Knight-Ridder in 1974), for $71.65 per tonne. For almost a century afterwards, this was the highest price ever paid for sulfur in the U.S. Ridder had purchased the sulfur for the paper mills that supplied his company. Unfortunately, the government never controlled sulfur shipments during the Spanish-American War, and sulfur pricing remained unchanged during the conflict.

After almost a century, not a lot had changed. In 1904, as the production of Union Sulfur began in earnest, their sulfur price was $18 per tonne, f.o.b. mine. In 1907, when there was initial friction with COISS, an "I'll-show-them-price" of $14 was established. When an agreement was reached the following year, the price returned to $18. During World War I, although demand soared, pricing was fixed at $22 per tonne for sulfuric acid manufacturers. Spot pricing to other markets did rise to $40. The pricing issue became a sore point between Union Sulfur and the U.S. government. The company was reluctant to accept this unfair, in their view, pricing. The government countered that the company was unpatriotic, accusing the company of having German sympathies, because of their founder. The government was so concerned about this issue that it was given as a reason for the War Industries Board to take over all sulfur supplies on July 9, 1918.

After World War I, pricing was expected to take a big hit, especially after the commissioning of the Big Hill Frasch mine in 1919. Then in 1920, there was a major economic downturn. The Frasch industry faced financial ruin. By 1920, sulfur stocks in the U.S. equaled five years supply! In the second half of 1919, the price had dropped to $15, where it remained until 1926. However, by then the situation had turned around. The 1920's were boom years, and sulfur demand soared. With increasing demand, the price returned to $18. Global pricing remained steady during the Depression, as a result of the cartel agreement between Sulexco and COISS. When UVZI replaced COISS, the Italians insisted that the international price of sulfur increase. The Americans, who were still leery of Rio Tinto and other pyrites producers, wanted it kept stable. The negotiations, as usual, went the way of Sulexco and a new deal was reached with UVZI in July 1934. Global pricing remained at the price set by Frasch in 1904!

The second biggest event of 1929 was the opening of the massive Boling dome sulfur mine. The curse of Texas Gulf Sulfur had struck again; for the second time (a downturn in the economy had hit after the opening of Big Hill in 1920), after opening a major Frasch mine, a recession hit, this time in record proportions. The Depression slashed sulfur demand. U.S. sulfur consumption had been 1.58 million tonnes, and Frasch exports were over 850,000 tonnes in 1929. Three years later, consumption and exports had dropped by more than 50%! From 1931 to 1932, elemental sulfur production plummeted from 2.1 million tonnes to 0.9 million tonnes, recovering to 1.4 million tonnes the following year, but still significantly below earlier highs.

One of the few signs of stability during the Depression, despite the massive drop in demand, was sulfur pricing! As with most industries, the 1930's were a lousy decade for the sulfur companies. However, unlike most, they survived relatively unscathed. The resilient margins of the Frasch companies were able to withstand such traumatic upset. By 1937, the industry had recovered and had established new production records, but demand began to decline again. Pricing finally dropped in 1938, the first time since 1926, from $18 to $16 per tonne, f.o.b. mine. Pricing stayed at this level until 1947, when it again returned to the level of 1904 at $18 per tonne. While the domestic price was at $18 per tonne, export pricing rose to $20 per tonne. During the 1930's and 1940's, Texas Gulf Sulfur was the largest sulfur company in the world, led by their massive mine at Boling dome, and Freeport Sulfur was second. Both companies made excellent returns at these levels. Between 1919 and 1943, the average profit-after-tax return for Texas Gulf Sulfur was 22% on capital employed. Freeport Sulfur was lower, but still respectable at 14% during the same 25-year period.

Before World War II, pyrites had been a governor on sulfur pricing, but after the war their influence diminished. More standard market fundamentals of supply-demand balance now influenced product pricing; a summary of which is given below:

1946 - 1956:	short
1957 - 1959:	balanced
1960 - 1963:	long
1964 - 1967:	short
1968 - 1976:	long
1977 - 1978:	balanced
1979 - 1980:	short
1981 - 1982:	long
1983 - 1986:	short
1987 - 1991:	balanced
1992 - 2006:	long

Within this sixty-one year time span, the market has been short for 21 years, in balance for 10 years, and long for 30 years; an interesting balance in its own right. The situation, though, is distorted by the several years where the market was short at the beginning and has been long at the end of this period.

At the start of World War II, the U.S. was producing more than 75% of the world's elemental sulfur. U.S. consumption reached record levels of more than 1.8 million tonnes in 1939. By the end of the war, the U.S. market was almost three million tonnes and U.S. production hit yet another new record in 1945. As the war wound down, there was fear of a collapse as had happened after World War I. In fact, just the opposite took place as demand soared even more, especially being driven by the superphosphate fertilizer market. Consumption in the U.S. hit 4.3 million tonnes in 1947! Adding to the growth of elemental sulfur was the decline of the pyrites industry of Europe, especially Spain and Norway. In the late 1940's, the strong market continued with demand outpacing supply, but the $18 per tonne price remained, and the export price went up to a modest $22 per tonne.

This amazing growth, though, was heading the industry towards a crisis. Between 1942 and 1950, demand exceeded supply. Stocks continued to drop and were approaching dangerously low levels, even though new production records continued to be set. G.W. Josephson and F.M. Barsigian began their **Sulfur and Pyrites** report in the U.S. Bureau of Mines, **Minerals Yearbook** of 1950 with the following:

For many years sulfur consumers have become accustomed to assuming that there would always be an ample supply available irrespective of demand. Therefore, it came as a real shock to the industrial world in 1950 when an acute shortage of sulfur developed.

The trend of the past decade had continued into 1950, which was not a surprise to anyone. The "shock" was the unexpected product allocation by the sulfur industry and the government. The reason was the claim that inventories had fallen below "safe working levels." Yet, the inventory level at the end of 1950 was still 2.7 million tonnes! While the levels were disconcerting, the actual inventories hardly seemed to be at crisis levels.

The years 1950 to 1952 were a period of unstable sulfur pricing. Domestic pricing first went to levels not seen since World War I, jumping 20% to $22 per tonne, and export pricing went up to a high of $27 per tonne. In the fall of 1950, the sulfur producers put their customers on allocation, which lasted until half way through the following year! Texas Gulf arbitrarily restricted shipments to 80% of contracted levels! Then the government jumped in. An unprecedented government rule was passed, *National Production Authority Order M-69*, to put quotas on domestic consumption, which restricted shipments to 1950 levels for the second half of 1951. The government-imposed allocation became more rigid in 1952, when shipments were limited to 90% of 1950 levels, even though stocks had risen the previous year (shades of Ferdinand and the TAC)! On July 17, 1952, Freeport Sulfur announced that the shortage was over, but M-69 was not revoked until November 5th. Oddly, a control order for sulfuric acid, M-94, had already been removed three months before on August 18th.

The U.S. Department of Commerce had also announced, on April 25, 1951, quotas on sulfur exports. Foreign governments protested. An international committee determined that there was a global shortage of one million tonnes of sulfur. Considering that before controls were in place, the total U.S. exports were less than 1.5 million tonnes, this shortage appears exaggerated. Nevertheless, the global sulfur industry was becoming so large that incremental increases were straining world trade networks. Governments around the world took control of allocating the limited sulfur supply to their most important industries. Some American politicians ranted that the U.S. was threatening its own industry by even exporting this much. On September 22, 1951, Representative Benton Franklin Jensen (1892 - 1970) of Iowa dramatically claimed that "the food supplies of American housewives" were being jeopardized! U.S. allocation control over exports of elemental sulfur did not end until March 1, 1953.

Yet, during this "crisis," almost three million tonnes of inventory lay untouched in the U.S. Gulf. The over-reaction to the situation which resulted in limited availability and higher pricing was demonstrated by rising inventories. In 1951, inventories increased by almost 200,000 tonnes, and the following year they increased again, this time by over 300,000 tonnes. How could this have been a crisis? Why did it take so long to relax regulations? The actions appear to be an overreaction to the situation. The yellow scare had been exaggerated because of the uncertainty of the Korean War.

Pricing was also regulated by the government during this period by the Office of Price Stabilization (OPS). The OPS established pricing in the range of $21 to $24 per tonne. Such regulations did not exist in other countries. The yellow scare saw panic pricing in some regions, peaking over $100 per tonne, with spot prices up to $200 per tonne being reported in early 1952. Sanity returned to the market place. On March 15, 1953, price controls were lifted by the government and pricing rose to $26 per tonne, f.o.b. mine, with export pricing about $3 per tonne higher. Delivered pricing to Canada was $40 per tonne.

While the profits of the sulfur companies soared, the shortage of the yellow scare era marked the beginning of the end of the U.S. Frasch industry. The crisis was a catalyst for the development of new sulfur sources. Oil and gas companies rushed to take advantage of the sulfur shortage; many recovered-sulfur projects were quickly approved, which until then had been languishing in corporate board rooms. Recovered sulfur from natural gas also began in Canada. By the end of the decade, Canada was a major player on the world scene, as was Mexico. Frasch mines opened in Mexico, the first outside of the U.S. World demand remained strong, but production charged past it. By the end of the decade, U.S. Frasch sulfur suddenly no longer dominated the world marketplace.

By the mid-1950's, the yellow bubble burst, exploded may be a better word. The sulfur world entered a new era as an international price war took place in the industry. Such aggressive pricing had not been seen since Frasch was going to show the Sicilians what his technology could do. This time, it was still a Frasch company upsetting the sulfur cart, but this time from Mexico. In early 1956, APSA had started the price war, with a $3 per tonne reduction to move its increasingly supplies of sulfur. In the fall of 1957, the Mexican firm had forced another $3 per tonne drop; the U.S. price was $24 per tonne, f.o.b. mine, while the Mexican price was the same f.o.b. Coatzacoalcos. Pricing continued to slowly weaken as U.S. producers reacted to shrinking market share, especially to Mexican producers. Sulfur was delivered to any major port in the world for ~$30 per tonne. The aggressive marketing stance of APSA lasted until early 1961.

In 1959, another prognosis in the **Canadian Minerals Yearbook**, C.M. Bartley reported that Canadian recovered sulfur production would reach four million short tons by 1970 (production was only 286,400 tonnes at the time!). He was wrong, but by only one year; this production number was reached one year earlier! He gave this prediction as a warning. Even though sulfur demand was steadily rising, global production, especially being spurred on by Canada, was growing at a much faster rate. There was a danger of a massive sulfur glut.

In the 1962 issue of the **Canadian Minerals Yearbook**, Bartley again wrote, this time about the changing of the sulfur guard:

> The history of sulfur has been one of the steadily increasing demand for industrial needs highlighted by occasional surpluses and shortages, price fluctuations, new production techniques, such as the Frasch process, and the rise and fall of whole industries, such as that of the Sicilian sulfur operation. Sulfur is again in turmoil but demand is still growing and Canada now appears to be in position to benefit.

His indirect warning was to the U.S. Frasch industry.

By the early 1960's, the U.S. Frasch companies were under siege, from Lacq in France; mined sulfur from Poland; and recovered sulfur from Alberta. Shipments out of Vancouver were especially gaining market share in the Pacific Rim area. Even domestic market share was being lost to Canadian and Mexican suppliers (and U.S. recovered sulfur producers). The costs of recovered sulfur were covered by natural gas revenue and thus its price was subsidized by the value of natural gas. The Mexican Frasch producers had lower labor costs and lower ocean-freight costs compared to the U.S. East Coast, since they were able to use non-U.S.-flag vessels. Under these new massive competitive threats, the U.S. Frasch producers did amazingly well. They were aided by rapidly growing global demand, and their new liquid sulfur logistics network.

Despite the globalization of the supply chain, the U.S. remained the dominant sulfur nation in the world, being the #1 producer, consumer, and exporter! The growth in U.S. demand between 1962 and 1966 was phenomenal, increasing from 6.3 million tonnes to 9.3 million tonnes. In 1964, Sulexco increased export prices (f.o.b. Gulf) from $25 per tonne to $27.50 per tonne. The following year, they increased again to $31 per tonne. Domestic pricing remained unchanged at $25 per tonne, until 1966, when increased to $28.50 per tonne. Strong demand led to record U.S. pricing of $38 per tonne, the following year, peaking at $40 per tonne in 1968. Some panic reminiscent of the yellow scare era took place. Suppliers in the U.S. and Canada restricted shipments. The high pricing sparked new mine openings and the reopening of some closed mines. In 1969, there were 21 Frasch mines operating in the U.S.

Canadian pricing was much lower. For overseas markets, they had to get the product to Vancouver, and remain competitive with ex. Gulf prices. In 1964, Canadian prices were $19 (Cdn.) per tonne, ex. works. During the sulfur shortage of the mid-'60's, pricing increased: $24 per tonne in 1966; $30 per tonne in 1967. In January 1969, pricing hit record levels in Canada, at $34 per tonne.

During these periods of tight supply dire warnings of a paradigm shift would be proclaimed. The penny-a-pound syndrome would never be seen again. For years to come, the situation would only get worse. How would the world cope? Corporations and entrepreneurs around the world sought to open new supply sources. Researchers reveled in the opportunity to explore the production of sulfur from "exotic" sources such as gypsum or coal. The upset soon passed, as another glut loomed over the horizon.

By the end of 1969, Canadian prices had plummeted to $13 per tonne! Under Cansulex (and non-members of the consortium, such as Shell, Amoco and Husky), Canadian sulfur sales expanded overseas, but not at the necessary pace to keep up with ballooning production. The global sulfur markets were deteriorating, as involuntary (and voluntary) sulfur production outpaced demand. In December 1970, representatives from Alberta met with their counterparts from other leading sulfur exporting nations, Mexico, France and Poland, in Mexico City to devise a strategy to reverse the trend; in other words, form a global cartel. These nations then attempted to implement inventory control measures. After less than a year, Alberta had withdrawn from the voluntary controls and the agreement collapsed. In 1971, the Canadian government organized other international meetings to attempt to deal with the issue, but without success.

The problem was that demand was fickle and production, oblivious to sulfur markets, just kept on increasing. There was some movement among Frasch producers. With the lower price, many Frasch mines in the U.S. closed. In 1971, the number was down to thirteen. While there was a major decline in the number of U.S. Frasch mines, there was only a nominal impact on production as the closures were generally smaller mines. Over 70% of the production came from the five largest mines.

In the 1970 issue of the **Canadian Minerals Yearbook**, P.R. Cote wrote:

> A new era in sulfur supply has begun. The effect of changes upon what we now consider to be "traditional" sources will be as profound on the global sulfur industry as was the advent of the Frasch process some fifty years ago.

The "new era" for sulfur was that the Frasch industry in America (i.e., "traditional") was no longer competitive to Frasch producers in other countries and the growing recovered sulfur supply even within their own nation. Overall, global production soared as Canada, Mexico, Poland, and France increased output. Poland was especially aggressive in world markets. Recovered sulfur was a yellow snow ball cascading down a mountain side growing larger and larger every year. Recovered sulfur pricing in the U.S. was more regionally influenced than Frasch. Usually, its pricing was a few dollars lower than the benchmark Frasch price.

Pricing slipped back to the infamous $18 per tonne. The situation was worse in Canada, when in December 1971, pricing hit a new low of $5.47 per tonne (Cdn.), ex. Alberta plant. Of course, the extreme market conditions again sparked fears of the end of the world as we know it. The U.S. Bureau of Mines reported (**Minerals Yearbook**, 1971):

> The oversupply situation was assessed as being basic in nature rather than cyclic, and was expected to continue over the short- and the long-term ranges.

The paranoia of paradigm shifts in the sulfur world is amusing. In less than five years, three such episodes were feared! None of them were related to fundamental changes in the industry. After the warning was given, true to form, the opposite soon took place. You would think that they would have learned by now. These extreme movements simply reflected the natural topsy-turvy world of the sulfur industry. A Frasch mine operated at full capacity or was shut down; there was not much room between the two. This feature forced temporary extremes between supply and demand. The growing recovered sulfur industry, oblivious to the rest of the sulfur world, was starting to make this separation worse.

In July 1970, the governor of Louisiana, Russell Long (1918 - 2003) stepped in to save his state's struggling industry. The protectionist bill "S4075" was to limit Canadian sulfur imports into the U.S., but it never passed. The government, though, was open to defend their sulfur interests. Freeport Sulfur and Duval Sulfur both brought dumping charges against Canada, and duties were imposed at the end of 1973. The charge was that Canadian sulfur was selling below production costs. While pricing was very low, the issue was controversial, since recovered sulfur was a by-product of natural gas production, and was argued to have no direct production costs at all. The cynicism of the U.S. action is demonstrated by the recovered sulfur producers in the U.S. not officially complaining of unfair pricing from the Canadians, for they knew better. The duties were dropped in 1978, but in 1981, Canadian sulfur suppliers were again found to be "dumping" product into the U.S. The Frasch industry was in trouble. This protectionist legislation was reminiscent of the feeble attempts of the Italian government to save the non-competitive Sicilian industry. The Frasch industry was simply no longer competitive; it was only a matter of time before the last mine closed.

Although still a factor, especially in the massive Florida market, the U.S. Frasch industry was no longer among global leaders. Frasch mines continued to close. Between 1968 and 1974, world demand of elemental sulfur soared, which provided a temporary reprieve to Texas Gulf Sulfur and Freeport Sulfur. U.S. Frasch production peaked in 1974 at almost eight million tonnes and then went into a decline. The Frasch industry was losing its competitive position, even within the U.S.:

1974 –   U.S. becomes net importer of elemental sulfur

1982 –   U.S recovered sulfur production surpassed that of Frasch for the first time.

1982 –   Poland surpassed the U.S. as the largest Frasch producer in the world.

1986 –   only four Frasch mines remained operating in the U.S.

1991 –   last year of major shipments to terminals in Europe

The entire Frasch industry would have shut down had there not been record high pricing in the late 1970's and 1980's.

A strange trend began in the sulfur industry: pricing started to steadily rise and quickly. At the end of 1973, the price of sulfur was at historic levels of $19 per tonne, f.o.b. mine; by the end of the following year, it had jumped to $31 per tonne, and this was just the beginning. These were turbulent times as there were record production levels and record demand. With both growing at similar rates, it was often difficult to know if the market was in balance or not. Adding to the confusion, infrastructure, especially logistics, could not keep up to escalating trade demands. This factor especially hampered Canadian shipments. At times, there were lots of product, but no means of getting it to the customers. The exception was the Frasch producers, whose global logistics network was well established. The end result was even though there was a serious sulfur glut, most of it was not available to world markets. Only the U.S. Frasch producers were in position to handle the demand, which pushed their system to its limits. While these factors contributed to the sulfur price inflation scare of the latter 1970's, there was more behind the incredible rise in price that had just started.

Frasch sulfur was energy intensive, and Frasch production costs (natural gas represented 20% of production costs) doubled during the Arab oil embargo in late 1973. The U.S. Frasch industry was consuming 50 billion cubic feet of natural gas per year. Ironically, the energy sector had become their number one competitor. The energy crisis forced sulfur pricing to record highs. The major industrial customer, the Florida phosphate industry, was expanding, and they were particularly dependent on Frasch sulfur. While recovered sulfur was a competitive option, no significant inventories were kept by U.S. oil and gas producers. Canadian producers were simply too far away. Only the Frasch industry had nearby major inventories to buffer fluctuations in demand by the volatile phosphate producers. The Frasch industry could have collapsed during this period, without price relief, and the major consumers were not yet willing to take a chance on that for now. The phosphate industry surrendered to higher pricing for security of supply, opening the pricing flood gates. Even though demand slackened during 1975, the average price jumped again to a new record of $65 per tonne at the end of the year! Canadian pricing lagged behind at $28 per tonne (Cdn.). Pricing then settled down, but only for a few years. Any hint of even a balance between supply and demand caused pricing to rise. In 1979, new records were established, as a price of $88 per tonne was quoted at the end of the year. At these prices, Canadian producers became interested in the Florida market; over 120,000 tonnes of sulfur were shipped from Canada to the distant Florida market, once the exclusive bastion of the Frasch industry. At the end of the decade, demand had finally surged past production and inventories dramatically fell in the U.S., but not total worldwide. The $100 barrier was quickly smashed (year-end Frasch price, f.o.b. mine): $119 per tonne in 1980; $139 per tonne in 1981. During the decade the pricing had increased over 700%! In 1980, the U.S. sulfur industry had net sales over one billion dollars. Canadian pricing lagged behind the Frasch pricing (Cdn.; ex. plant): 1980 - $72 per tonne; 1981 - $63 per tonne, North America; $82 per tonne, overseas. Sales in the Canadian industry in 1981 were over $800 million.

While pricing reached the highest levels in the history of the industry, inventory levels were also shattering records, both in the U.S. and, especially, Canada. North American inventories were 26 million tonnes by 1977! Market fundamentals had fallen apart, again. This time, though, there was method behind the madness. The sulfur blocks had become an integral part of the pricing strategy of Canadian producers. Originally, sulfur had gone to block only when no one would buy the sulfur, at any (reasonable) price. Now, it went to block when no one was willing to pay the price for the sulfur. The blocking of sulfur directly controlled inventory and indirectly controlled global pricing. In 1982, B.W. Boyd, in the **Canadian Mineral Yearbook**, wrote:

> However, as demonstrated during the 1970's, the price of sulfur is inelastic over a wide range of prices. Therefore, if the Alberta producers dispose of their reserve stocks as quickly as possible, prices would fall and, the reserve stocks would be sold at less than the long-term price trend. Thus the reserve stocks could be more responsibly managed by liquidating them when the price rises above the long-term trend and by building them up when prices fall.

Logistics was also a governor on sulfur availability from the blocks. During these periods of strong demand, there was not nearly enough capacity either to melt the blocks or to ship the extra volumes out of Vancouver. In other words, it was much easier to put the sulfur into blocks than to take it out.

Pricing finally stabilized during the recession of 1982 but considering the huge drop in sulfur demand, the price would have been expected to decline dramatically. The price at the end of 1982 remained above $100 per tonne. In 1985, Canadian pricing also reached this level in Canadian dollars, but the U.S. Frasch price had risen back to $123 per tonne. For most of the decade the price was erratic but at relatively high levels, and only started to seriously decline in the second half of 1991. The decline came as a surprise to the industry as global production had been curtailed by the Gulf War, and other factors. In this mixed-up industry, global inventories, especially in Canada, had plummeted and prices were in free fall.

The producers never had a period like the good times of 1980 to 1990, when pricing never dropped below $75 per tonne, f.o.b. mine. The high pricing had postponed the demise of the U.S. Frasch industry. However, its cost structure, especially with high energy costs, made it difficult for most producers to remain profitable if pricing dropped below $100 per tonne. By this time, the Frasch industry had declining influence on the global industry. The baton had passed to recovered sulfur, especially from Canada.

Throughout this period, the benchmark sulfur price had been the f.o.b. Frasch price at their Tampa Bay terminals. In 1992, this practice was abandoned. The f.o.b. Vancouver price (in U.S. funds) became the leading price indicator. Recovered sulfur had been sold at 20% to 50% lower than Frasch. Earlier, their influence on global pricing had been minimal, as this sulfur was not readily available for export. With the opening up of Vancouver to export shipments and then an efficient system to get it there through Sultran, Canadian recovered sulfur producers set the global standard. For now the price was maintained less than Frasch, as recovered producers needed more of their market share.

An ominous turning point for the industry took place in March 1992, when Canadian inventories increased for the first time in the past dozen years. Global elemental sulfur production started to rise at a considerably higher rate than the previous two decades. The rise in production took pricing down with it! The new Canadian marketing organization, Prism took control of export sales at the beginning of 1992, and was thrown into a hornet's nest. Aggressive marketing saw exports increase, especially to the U.S. The price f.o.b. Vancouver started in January at $70 per tonne (U.S.), after a major drop the previous year, and had dropped to the mid-$40's per tonne by the end of the year. Overall pricing (f.o.b. Vancouver) had dropped from $100 per tonne at the beginning of 1991 to $30 per tonne three years later! Dumping charges were again laid against Canadian producers, being initiated by Pennzoil Sulfur (ex. Duval) in late 1992, 1993, and 1994. In July 1995, duties were again assessed by the U.S. Department of Commerce. After almost three decades of protectionism, the duties were lifted one more time by the ITC on January 1, 2000. By then, there was no Frasch industry left to protect. By the end of 1993, liquid pricing had fallen to $0 per tonne, f.o.b. Alberta plant! As prices declined, Prism attempted to steady pricing by again building inventories.

The traditional supply-demand drivers of commodity pricing were relatively dysfunctional in the global sulfur industry by this time, even more so than normal! While demand was still influenced by market fundamentals, sulfur supply had become increasingly distant from demand, as recovered sulfur became more and more the dominant source. In today's industry, sulfur supply is driven by demand, not demand for sulfur but for oil and gas. The prime directive of sulfur businesses within the global oil and gas industry was to manage the fatal sulfur production, as economically as possible, while never allowing it to interfere with the core energy business. To the oil and gas industry, sulfur is not an asset but a threat to their prime core business. Demand for sulfur was relegated to only a secondary factor at best, as the sulfur business was becoming more a disposal unit than a supplier. This intrinsic feature of the recovered sulfur industry created a strange business environment with supply growth being relatively fixed and independent of variable demand. The situation was a natural outcome of the circumstances. Recovered sulfur demand and sulfur supply are disjointed and would only fall into balance serendipitously. Considering the collapse of traditional market principals, the overall sulfur industry has been surprisingly stable.

As yet another century began, the price of sulfur was off from historical levels; it was lower! U.S. sulfur pricing hit $10 per tonne, a price last seen in 1904! At this time, the Vancouver price was $18 per tonne (U.S.). Increasing Chinese demand for its rapidly expanding phosphate capacity has helped bring back the level to the penny-a-pound benchmark. The U.S. sulfur pricing has since rebounded to the $30 per tonne range, while Vancouver was at $63 per tonne (U.S.) for 2004 and 2005. In the latter half of 2006, pricing again started to erode, reaching $47 per tonne in late summer.

The abnormal historic pricing of sulfur is incredible. For over two centuries, the "normal" pricing had been about a penny a pound. Had the pricing been in constant dollars, the situation would have not been so absurd. Since it is actual dollars, the pricing in constant dollars has plummeted over the decades. A $0.010 per pound price today would be equivalent to $0.065 in 1950 or $0.200 in 1900. Over the century, the market price for sulfur has declined in constant dollars by about 95%!

Now into the 21st century, sulfur pricing has returned to the traditional levels of a penny a pound. A daring statistician would project that the price would be at this level in the next century as well.

Chapter 6

Future Sulfur — Oil Sands

As Williams Haynes stated in 1959, in his book, **Brimstone, The Stone that Burns**:

> Indeed, though chemical forecasting is dangerously apt to back-fire, it is now safe to say that as long as man inhabits this planet he will suffer no lack of sulfur.

As we gaze into the yellow crystal ball to forecast the future of the sulfur industry, the past is the fundamental place to start. The adage, it is easier to forecast the past than the future, has special applicability to sulfur.

At whatever rate of growth, the critical issue is will demand keep up to supply? As has been the case for the past century, the next century demand for sulfur will depend on phosphate fertilizers. The insatiable requirement to replenish the nutrients from the soil that agriculture "mines" from the earth will likely maintain a strong core demand for sulfuric acid and its surrogate sulfur. A change, though, in regional distribution is possible, even likely. Can Florida maintain its dominant position against increasing capacity in other nations, especially China? If a phosphate migration takes place, the U.S. could soon find itself to be a net exporter of sulfur once again.

Where is the future production of elemental sulfur headed? On present growth trends, different projections can be made. A sharp decline in the growth rate of elemental sulfur production took place after 1974. If the trend since is taken, global sulfur production will edge up only slowly and reach the fifty million tonne mark in 2032. However, for the past dozen years, sulfur production has been increasing at an amazingly steady and faster pace; if this trend continues the fifty million tonne mark will be reached twenty years earlier. The recent growth spurt is being led by the major fields at Caroline, Astrakhan and Tengiz. The prognostic question is, can this pace be sustained? If so, where will the sulfur come from?

What will be the source of sulfur in a decade or a century? Without doubt, recovered sulfur will still dominate world supply for, at least, the next fifty years. The question is recovered from where? Will the future supply of sulfur originate in the northern Caspian Sea or the Persian Gulf, with their massive oil and gas reserves? Although new fields may still be found in these regions, the most likely source to replenish the global appetite for the yellow element is still Alberta, in particular Northeastern Alberta. Alberta will retain its global position for decades to come, not from future undiscovered sour gas deposits, but from the oil sands region. The monster projects that have been announced will significantly increase the world's supply of sulfur (see Table 6.1). The timing of such mega-projects are often delayed, or cancelled all together. However, if only a minority of these behemoths go forward, production growth at this accelerated pace will be maintained. The oil sands contain the largest sulfur reserves in the world at two billion tonnes.

Table 6.1. Oil sands – projects

Projects	Companies
Alberta Heartland (*)	BA Energy
Aurora (*)	Syncrude
Christina Lake	EnCana
Christina Lake	MEG Energy (17% CNOOC)
Dover	Devon Energy
Firebag	Suncor
Fort Hills (*)	Petro-Canada/UTS/Teck Cominco
Fort MacKay	Fort MacKay First Nation
Foster Creek	EnCana
Great Divide	Connacher Oil & Gas
Hangingstone	Japan Canada Oil Sands/ PetroCanada/Nexen/Imperial Oil
Hilda Lake	Blackrock Ventures
Horizon (*)	Canadian Natural Resources
Jackfish	Devon Energy
Jackfish 2	Devon Energy
Jackpine	AOSP (Shell)
Joslyn Oil Sands Project	Total SA/Energy Plus Refining
Kai Kos Dehseh (*)	North American Oil Sands Corp.
Kearl Lake Oil Sands Project	Imperial Oil
Long Lake (*)	Opti Canada/Nexen
MacKay River (expansion)	Petro-Canada
Mahihkan North	Imperial Oil
Meadow Creek	Petro-Canada/Nexen
Muskeg River (expansion)	AOSP (Shell)
Nabiye (expansion)	Imperial Oil
Northern Lights (*)	Synenco Energy/Sinopec

Table 6.1. Oil sands – projects

Projects	Companies
Orion	Blackrock Energy
Peace River	Shell
Primrose	Canadian Natural Resources
Steepbank (expansion)	Suncor
Strathcona (*)	PetroCanada
Sturgeon (*)	North West Upgrading
Sunrise Thermal	Husky Energy
Surmont	ConocoPhillips/Total/Devon
Tucker Lake	Husky Energy
Voyageur (*)	Suncor
Whitesands	Petrobank

(*) heavy oil up-grader planned

6.1 SULFUR & THE OIL SANDS

The oil sands are a geological formation of a mixture of sand, water, clay and bitumen, which is a tar-like, thick heavy oil, with an average sulfur content of 4.8%. Deposits are found in Northern Alberta, at Athabasca (near Fort McMurray, includes Wabasca deposits), Cold Lake (NE of Edmonton), and Peace River (NE of Grand Prairie). The massive proven reserves of the oil sands are 175 billion barrels, with potential reserves exceeding 310 billion barrels. Among proven reserves, the oil sands are only second to Saudi Arabia.

After the heavy oil has been separated from the sand, the bitumen cannot be processed in a normal oil refinery. It must be treated in an upgrader, where the hydrocarbon macromolecules are broken into smaller fragments. There are two processes commonly used. Petroleum coke can be produced, where any sulfur components remain in the final product. The more common process involves hydrogen cracking, which produces synthetic crude oil, which is then further refined in traditional oil refineries. The sulfur trapped in the organic matrix of the oil is converted into hydrogen sulfide. A traditional Claus process can then be used to produce sulfur.

The issue of sulfur for the oil sands had even been discussed in the "Blair Report." The sulfur reserves were estimated to be between one billion to three billion tonnes. In 1955, T.H. Janes reported on the sulfur potential of the oil sands (**Canadian Mineral Yearbook**):

> A huge potential source of sulfur in Canada exists in the exten-
> sive deposits of bituminous sands along the Athabasca River in
> Northern Alberta...It is estimated that a 20,000-barrel-per-day
> bitumen plant would result in the recovery of about 140 tonnes
> of sulfur.

In 1973, G.H.K. Pearse, writing in the **Canadian Minerals Yearbook**, projected that sulfur production from the oil sands could reach five million tonnes per year by 2000. However, these projections were drastically curtailed in subsequent reports. While it is easy to announce oil-sands projects, the massive financial commitment and fickle oil pricing often delays, or cancels, their implementation. A common deterrent is escalating costs between the planning of the project and construction date. By 2000, only Suncor and Syncrude were operating sulfur plants in the oil sands.

The first sulfur production out of the oil sands was by Suncor: 1968 - 23,000 tonnes; 1969 - 23,400 tonnes; 1970 - 47,000 tonnes. The first sulfur shipments out of the region, though, did not begin until October 1974. Exports of 30,000 tonnes were sent to Europe via Quebec City. At the end of 1974, the block inventory already stood at 375,000 tonnes. While logistics had been a major obstacle to the sour gas plants in Alberta, the situation was far worse for the oil sands region. Sulfur production surpassed 100,000 tonnes from the oil sands for the first time in 1976. Shipments of sulfur, though, remained low through the 1970's, with most of production going to block. Afterwards, shipments rose from Suncor, with most products moving by truck to the nearby rail connection at Lynton, AB. Unlike Suncor, Syncrude has blocked their sulfur. In 2005, their sulfur blocks are over six million tonnes and are one of the largest site inventories of sulfur in the world. The combined sulfur production from Syncrude and Suncor was 1.2 million tonnes in 2005. By 2015, oil sands production will almost triple to three million barrels per day, and with it sulfur production should be over three million tonnes per year. In 2004, the National Energy Board projected in their **Canada's Oil Sands, Opportunities and Challenges to 2015: An Update** that sulfur production from the oil sands alone could hit 10 million tonnes by 2030.

6.2 BEYOND THE OIL SANDS

Someday, in the next century or two, the recovered sulfur industry including the oil sands, too, will pass. Whenever this takes place, where will the sulfur come from? Huge reserves of pyrites are still available in various locations of the world. The past creation of the sulfur-pyrites industry had been somewhat serendipitous. Imagine if they had been twice lucky. Although vast supplies of pyrites still exist, they have been uneconomic due to the historically low price of sulfur, especially recovered sulfur. What if recovered sulfur had not developed on such a massive scale after World War II? As Frasch production became more expensive and reserves depleted, pyrites would have become the leading source of the yellow element today. This "what if" scenario was not an ivory-tower discussion. The Frasch industry took it very seriously. In their competitive analysis of the global sulfur industry, pyrites were their greatest competitive threat until World War II. Their pricing strategy acted as a deterrent to the growth of pyrites. Is the prominent position of the pyrites industry in the sulfur world truly over? Or has it only gone into hibernation? When the day comes that recovered sulfur from oil and gas is no longer available, pyrites may indeed once again become a powerful source in the sulfur world. With its inherent drawbacks, the rebirth of pyrites is a long way off, but if recovered sulfur ever falters, they are waiting in the wings like an iron phoenix.

One of the largest untapped sulfur sources is gypsum. In 1967, Elcor Corporation, through its subsidiary, National Sulfur Company, planned to invest in a plant in Rocky House, TX, to produce sulfur from this source. The project was abandoned in 1970, because of weakening sulfur markets and technical challenges. The time of gypsum as a major sulfur source has not yet arrived, nor should it be expected in the coming decades. Perhaps, the infamous Bearberry "sulfur" well will be reopened northwest of Calgary. Coal, where sulfur is found as a complex mixture of organic compounds, is yet another untapped source of sulfur for some future decade or century.

The real issue is not availability but economics, as P.R. Cote and W.E. Koepke reported in the 1967 issue of the **Canadian Minerals Yearbook**:

> The sulfur shortage is not due to a lack of sulfur within the
> earth's crust but rather to a problem of economic recovery.

As the world has become accustomed to sulfur at a penny a pound, the pivotal question is can a source be found to keep the price at this historical level for the remainder of the present century, or even the next?

THE SULFUR ENTREPRENEUR

7.1 THE DESPERATE ENTREPRENEUR & THE RELUCTANT INVESTOR

The evolution of a global commodity industry, such as sulfur and sulfuric acid, follows, in its simplest form, only five basic stages:

I. Ideas are the bacteria of industrial evolution;
II. Experimentation spawns a plethora of simple creations – the first evolutionary step is taken by scientists and inventors who take the initiative and prove on a small-scale that the idea works and has financial potential; most disappear and are only seen again as technical fossils within scientific journals, but a few are transformed by entrepreneurs into something much greater;
III. Commercialization is the evolution from the simple laboratory creations into plants – a prototype plant is championed by determined risk-takers, whom some call entrepreneurs; this group must overcome what appears to be endless technical problems to reach design parameters; meanwhile, the business produces one thing, mounting debt;
IV. Financial sustainability separates the annuals from the perennials – the business to survive must generate enough cash flow to cover all costs and produce a reasonable return; non-technical factors, such as cost-control, purchasing, logistics and, especially, marketing, become increasingly important;
V. World-scale production is the top of the industrial evolutionary scale; corporate visionaries now take-over; investments are made in huge plants and/or plants around the world; marketing, especially, is pushed to its greatest limits. However, are they dinosaurs or mammals? Even a mammal can be squashed under the foot of a dinosaur.

An antiquated adage of success is one-in-ten between each stage, but the reality may be better reflected in a more logarithmic series, in other words, each world-scale industry takes ten trillion good ideas! Unfortunately, serendipity by being at the right place at the right time often separates success from failure (Leblanc had found this out the hard way). Careers and fortunes are made and lost, despite the brilliance, or lack of it, by those involved.

While often as creative in their own right as any scientist, entrepreneurs separate themselves from inventors, by taking serious risk, sometimes everything. The odds are against them as most fail. These forgotten business heroes are those who had the insight to see a great potential in the innovation, often oblivious to the inventors, and who overcame the endless technical and financial hurdles. The industrial accolades for the Sulfur Age belong to the likes of the Muspratts and the Tennants, who hold, at least if not more so, equal status to Roebuk and Leblanc. These business pioneers did not discover the technology; they did not even build the first plants (they were even a decade behind the industry that had developed in France). They did take an industry to where no one even dreamt it could go!

The evolution of a global chemical industry is well illustrated by the first such industry, sulfuric acid. Glauber had introduced a technology that stalled for many decades. The acid was simply too expensive for any widespread development. Roebuck changed that. A technology, though, without a market does not an industry make. Such was the situation with the discovery of the Chamber process in 1746. While important to the cottage industry at the time, a true industry did not develop until after the Leblanc plants began using the Chamber process in the early 19th century. Even Leblanc himself, while being a catalyst, also did not directly create this industry. If the industry was dependent upon only his development, it never would have happened.

The stages of the evolution of industry are on a slippery, steep slope, requiring deep pockets of patient investors. An inventor needs an investor for his invention to become a venture. Leblanc was luckier than most entrepreneurs; he already had a financial backer, the Duke of Orleans. The Duke had banked his scientific studies and the building of the small plant. Unfortunately, the Duke went down with the Revolution and took Leblanc's plant with him. The Duke was Leblanc's "government" investor. In venture projects, there is a myriad of investors and pseudo-investors, and a few types of entrepreneurs:

7.1.1 INSTITUTIONAL INVESTORS

- *Government:* great source if available without too much red tape.
- *Venture fund:* seldom does a new venture, with an unproven technology or poor balance sheet, meet their guidelines.
- *Vulture fund:* similar to the above, but to attract the financing the entrepreneur has to virtually (or totally) give up control of their company; many vulture funds will "watch" struggling venture firms to see if an opportunity for them pops up that they can take advantage of; an antiquated term for this group was carpetbaggers; Baruch was more this type of investor in his dealings with Einstein.
- *White knight:* institutional fund investors, who invest in a high-risk venture for special reasons of their own (a dream come true for the entrepreneur, but they are few and far between); an example of this type of investor was Swenson with Freeport Sulfur.

7.1.2 PRIVATE INVESTORS

- *Angel*: a group of private investors, who are supposed to be more open to riskier, high potential investments, but often develop criteria similar to institutional firms.
- *Sophisticated private*: this snobbish title implies that they are balance-sheet driven and not future cash-flow driven; in other words, they are less likely to invest in new technology ventures.
- *Gamblers*: individual investors who are driven more by the business plan than the balance sheet; often they are from the networking of the entrepreneur, and are the heart-and-soul of many new venture projects.
- *Commissioner*: a common source to attract private investment, but part of the investment goes to the middleman as a commission, often ranging from 5% to 15%; be careful of those who demand up-front fees to attract funding, with no funds of their own and no commitment to actually come up with any; desperate entrepreneurs may turn to such sources, only to find they are more in debt than they were before.
- *Window-shopper*: these so-called investors have no source of funds, they know the proverbial someone who has money to invest; they are only looking to find a project to attract investors into a new portfolio or to form their own commissioner company.

7.1.3 ENTREPRENEURS

- *Swindler*: There are two sub-categories within this type. The first is the unscrupulous promoter who exaggerates the value of their project with the intent to defraud their investors. The second is the desperate promoter who exaggerates the value of the project to keep their dreams alive at any cost. In the sulfur world, Frasch had accused some of the earlier projects at Calcasieu as being in this group.
- *Dreamer*: While having a good idea and honest intent, the dreamer rushes blindly into complex projects and is ill-prepared for the technical and financial burdens that lie ahead. With a hope-and-a-prayer attitude, they can succeed, but more luck than skill is usually required. Among the early sulfur entrepreneurs, Einstein may be accused of falling into this category.
- *Professional*: the professional has been careful to secure proper financing and conducted a proper evaluation of the technology. The best example of this group of entrepreneurs is Frasch himself. Even "professional" entrepreneurs have some of the character of the "dreamer," and may be tinged by a bit of the "swindler."

At the outset, the novice entrepreneur believes in a financial fantasy world. An idealistic example of a meeting between entrepreneur and investor was presented by the *New York Times* on January 10, 1887 (p. 2). The promoter begins by asking if the financier wants to make millions. He then goes on to prove his point, and the financier agrees. While seldom do such meetings ever go so easily that was not the intent of the article. The late 1880's had been a turning point in the commercial relationship between the South and the North. After the Civil War, the investors from the North were known as carpetbaggers, who had come to exploit the troubled times of the South. By 1887, Southern entrepreneurs were now going to Wall Street to get financing for their projects. The Wall Street capitalists are not mentioned, but they could very well have been members of the ASC. The example cited in the article was an investment in Louisiana sulfur!

While the entrepreneur dreams of financiers beating down his door, this is a venture fable. Most meetings do not end with such an idyllic conclusion as above. Financiers, with an overtone of irritation, claim that they will get back to them, which they never do. When contacted, the financiers are just completing another deal, or waiting for new financing to come in. Seldom does the entrepreneur ever hear that there will be no investment; the opportunity silently fades away. As time passes, dreams start turning to nightmares, as the entrepreneur must press harder and be willing to sacrifice more. Their own livelihoods are often put at risk so that their project becomes a reality. This process is not for the faint of heart.

The relationship between the entrepreneur and the investor is convoluted. Generally, the two are antipodes of one another, one daring and passionate, the other reserved and analytical. There is also a natural distrust between the two that goes beyond their personalities. The entrepreneur eyes the investor as a potential snake that may try to steal their brilliant plans or lucrative business. He can become intolerant and rude. He can't understand (yet once again) why the investor is too stupid not to see how great the project is. The investor is leery that the swashbuckler entrepreneur is a pirate out to waste, or even steal, his money. He becomes condescending and skeptical towards the presentation and the presenter. Is the project all smoke-and-mirrors, driven more by passion than sound financial judgment? As a "polite" gesture the entrepreneur is taught the realities of the investment world. The blood boils inside the entrepreneur as yet again he must listen to this financial malarkey.

The cases of pure swindles are exaggerated and have been unfairly extended to all failed projects and taint most new honest ones. The early mining efforts at Calcasieu, for example, saw the fall of one investment group after another. Frasch, himself, remarked on the failure of these projects. He had even met with many of them [*J. Ind. Eng. Chem.*, **4** (2), p. 136, 137 (1912)], "people who told [him] of having lost money in the various schemes…" Frasch also complained about these companies "telling the prospective purchaser that his fortune was made if he but owned a few shares of the stock of that particular company." He even accused them of "hiding" any negative aspects. What did he really expect them to do? Did Frasch, himself, not promise his investors "fortunes?" What Frasch had originally "promised" his investors is not known, but he surely would not have told them that several years would pass before the company would be cash positive. Despite the criticism from Frasch, the potential of Calcasieu was real. The technology of the times, though, was not up to the task.

Despite these challenges, investors are eventually brought into the venture. At last an investor is willing to risk their funds, or those of their clients. They are brought together by the sirens of the elusive financial rewards that await them if they are successful. The entrepreneur is thrilled that the standard start-up investment has been raised. Now, he can build the plant, cover initial operating costs and accounts receivable bridging, until operations are projected to become cash positive. Even the investor is a little bit excited about the prospects, but a little more nervous about the money. The reluctant partners are now stuck with each other, as the real problems manifest themselves!

Oops, a small mistake has taken place in the perfect plans of the venture: the initial investment funds have run out. Capital costs were more than planned, and/ or initial funding was less (and later) than budgeted. Unfortunately, most entrepreneurial ventures ultimately require a second phase of investment, much to the chagrin of the entrepreneur and the initial investors. The difficulties in acquiring the new financing exceed that of the first. The mystique of the new project is now tarnished by a deteriorating balance sheet. Even though the final goal is closer, doubt rises among present shareholders. There is increasing fear of the toilet-bowl-effect, where investment funds are being flushed away just to keep the project alive and not progressing towards financial sustainability. Through these endless times, the company produces one thing, debt. The negative cash flow extends beyond everyone's wildest fears. Costs are escalating with no cash flow coming in. The project is going into bankruptcy. This vulnerable transition takes down many exciting chemical discoveries and with it many entrepreneurial firms.

The initial investors are becoming increasingly ensnared in a common trap. More funding is required, and new outside investors are difficult to find. The proverbial choice is to walk away and lose their investment or invest even more to keep the company alive. If they choose the latter, there is even more pressure upon them if further funds are required. The enticement is that one more dollar might be all that was needed to put the company over the top; so the entrepreneur keeps claiming. They are badgered by this perpetual optimist to support the struggling company. The entrepreneur's onslaught on them is never-ending. They now wished that they never had got involved. How could they have been so stupid? They search for a scapegoat. It was not really their fault. It was that damned entrepreneur; he had promised them...

Regardless of the wording, the entrepreneur can not promise the investor anything. Granted the results will not come as fast as projected (not promised!) by the entrepreneur. On the other hand, the investor can not claim with any sincerity that they really expected the financial results within the time frame proposed. The shareholders are not "innocent" victims of the unscrupulous promoter. They had voluntarily gotten themselves into this mess. Investors like to complain that they were misled by the entrepreneur, but they really knew better. They have only themselves to blame. They tried to hit a financial home run. Many strike out. Some get more than they hoped for. In both cases, no one forced them to play the game.

What had gone wrong? The financial plan had "proved" positive cash flow within six months. The sophisticated financial model, of course, is usually no more than a prettied spreadsheet. Many of the inputs are estimates. With such a system, there is much flexibility on the determination of these numbers. However, the investor had checked these and found them to be sound. The much maligned financial model still shows that they all will become rich beyond their dreams if the project succeeds. What then has gone wrong? The strength of the model is the financial outcome; the weakness is the timing. The poor judgment of timing is not from a lack of technical or financial skills or a deliberate attempt to deceive the investors but from the endless problems: the broken pieces of equipment, the necessary improvements, the metallurgy, the quality problems, the reliability issues, availability of raw materials, the lack of skilled tradesmen, delays in environmental permitting, the market resistance... The list goes on and on.

So far, the analysis of entrepreneurial ventures appears to be directing the investor away from such projects. Not so, the above is only a warning to investors to be prepared to face such obstacles. While there may be high risk, there is higher reward. Only those that have the stamina and courage will reap the benefits. Even the revolutionary Frasch technology had become ensnared in the fatal trap that captures many innovations: the vulnerable period between building the first prototype operation and bringing it up to design specifications. The early years of Union Sulfur were typical of the struggle between entrepreneur and investor. The company had formed in early 1896 and by the end of 1897, it was almost bankrupt, a natural state of new-technology companies within 24 months of start-up. By September 1897, the new company had a mortgage for $125,000, plus $100,000 in bills payable, $72,000 in other expenses and $21,000 in interest payments. With this debt of $320,000, Union Sulfur had sales of only $39,000! While steady progress had been made on the technical side, production was still erratic, and the energy costs of the hot water were so high that the process had no chance of being commercially viable. The influential *Engineering & Mining Journal* wrote in their review of the year 1898: "It is now conceded that this process is a complete commercial failure." The article had been written by the editor, Richard Rothwell, with some cynical glee. The Canadian-born Rothwell knew the site well, for he had earlier been a partner in ASC. Under the circumstances, what "sophisticated" investor in their right mind would have stayed with Union Sulfur? After years of growing debt, waning support from investors, and skepticism among experts in the field, this exciting new process appeared to be destined to the write-off junkyard of "bleeding-edge" technology, resulting in commercial ruin to many, a tax write-off for the more well to do. Yet, Frasch had a brilliant technology that revolutionized an industry, but he, too, had been caught by the time albatross.

The initial shareholders were in a bind. Even though, they were generally a wealthy bunch, in the back of their minds, they must have thought they were throwing even more money away. Under the circumstances the prudent action would have been to cut their losses and run, and some did. Frasch had been fortunate to come across this particular group of financiers. The entrepreneurial spirit also resided within these investors as well, for they had formed ASC earlier. Spurred on by Frasch's dogged determination (and huge returns if the process ever worked to design), his badgered partners reluctantly agreed to provide more financing. Perhaps, the investment partners, themselves, felt trapped. After pouring significant funds into the project, could they simply walk away and lose it all? But for how much longer could they keep putting their money into this apparently bottomless pit, more properly a mine?

The new investment did little to turn the fortunes of Union Sulfur. More years passed, and still no sustainable production. Even after the oil discovery at Spindletop, profitability seemed distant. The action of the savvy Twombly to try to sell his shares in the middle of 1901 is a stunning example of the dichotomy of the new-technology investor. During the summer, he had tried to bail out, offering his shares to Captain Lucas at a nominal price. Luckily for Twombly, Lucas refused. A common deception in such entrepreneurial situations is that even when truly close to success, judgment is clouded by the reality of the past. Was it truly any different this time? In this case, it was. Without doubt, the other investors in 1901 would have understood Twombly's actions and may have joined him if he had been more successful. A survey would have likely shown that if they had to do it all over again, they would have never invested in Calcasieu. An exception would have been Frasch, for it was his dream after all.

New ventures often succeed (eventually) if properly financed and provide handsome returns to their investors; Frasch and Union Sulfur are an entrepreneurial success story. The financial support against a balance sheet that defied rational investment paid off handsomely for all concerned. After the fact, it was a "brilliant" investment. For many years, dividends were 100% per month. They all became rich (actually, in this case, more rich than they had been before).

Another illustrative case of the relationship between the entrepreneur and the financial world was exhibited at Big Hill (Texas Gulf Sulfur). In this case, Einstein had been the entrepreneur to promote the project. The St. Louis businessman and his syndicate raised enough basic funds to obtain rights to the sulfur at the Texas site, but couldn't attract the investment necessary to build the mine. They were stuck in an investment no-man's land. Their new company was floundering. Einstein, though, did not give up. Baruch had been the one major financier to show an interest in the budding sulfur business [a textbook example of the real world of the entrepreneur and the venture capitalist is outlined in the biography **Bernard M. Baruch, The Adventures of a Wall Street Legend** by James Grant].

In 1909, Baruch had first become involved in Freeport Sulfur at Bryanmound, after being asked to look into the investment potential by J.P. Morgan & Company. The financial specialist brought in the noted mining engineer Seeley Mudd and Spencer Browne to provide a technical assessment of the property. Baruch knew not to solely depend upon the appraisals by the original entrepreneurs of the site. There is often some bias in the appraisals for the financier as well, but usually on the negative side. The appraisal by the investor group is designed to protect the financier from making a bad investment. Neither appraisal is "wrong." Since they are estimates, there is a choice in many input parameters in the financial analysis. The choice is naturally influenced by the position of those inputting the data. Oddly enough, a balanced viewpoint is difficult to achieve; most are optimistic or pessimistic. In this case, Mudd concluded that the odds were less than 50/50, not the most encouraging news. Baruch, though, was willing to put up some of the money, but J.P. Morgan was not. Morgan had made a serious mistake, as Bryanmound became a productive sulfur mine and Freeport Sulfur a successful company. The entrepreneurs at Bryanmound were luckier than most. They found a major financial backer, who had personnel ties to the region and so was less cautious, Eric Swenson, the director of the First National City Bank of New York. Without such a "white knight" investor, Freeport Sulfur would have likely struggled as much as Texas Gulf Sulfur.

Browne, while investigating Bryanmound for Baruch, had heard of Einstein's venture in nearby Big Hill. He approached Einstein to check out the other potential sulfur mine. He found it to have, at least, the same potential as Bryanmound. Einstein must have been elated that such an esteemed investor as Baruch had shown interest in his project. However, his enthusiasm was soon dampened. Baruch did not respond, at first, to his queries. Einstein continued to make proposals to entice Baruch to invest, but was turned down each time. The interest of someone of the caliber of Baruch throws the entrepreneur into ecstasy, for he sees the financial savior has arrived. However, the financial prophet may be the devil in disguise. Baruch was coy and cautious, the worse traits for the entrepreneur desperate for money. Yet, Einstein, as is often the case, had little choice. Interest from someone with money was better than no interest at all. If he had another option, he would have grabbed it. Baruch controlled the situation. All Einstein could do was to nudge him forward. Einstein tried to be more and more creative in the offers to entice Baruch. Silence is an effective negotiating ploy, especially when the deal is more important to the other party than it is to you. Einstein and the other original shareholders were negotiating with themselves, a common mistake. Baruch remained silent; why not, since he kept getting better offers. Finally, in June 1912, money from Baruch slowly began to appear. The funds were not enough for construction to begin. The company was still floundering. In November, Bryanmound had started up, and Big Hill remained just a big hill. The delays in start-up put the company into greater and greater financial hardship. Although there were no operations yet, and thus no cash flow, there were some expenses that had to be paid. With no income, the source of funds had to be investments or loans. Baruch reluctantly provided more funding, but with a common restriction; his loan was to have first rank. The entire enterprise was at a crossroads. Both sides were becoming more cynical of the other's position and intention. The entrepreneurs increasingly felt that the potential investor was just using them, deliberately dragging out the discussions so that the latter can just take over their wonderful opportunity. On the other hand, the financier wonders if the idea is so good, why doesn't anyone else come to the table? Secretly, he hopes that someone else will take the initial step and if they are successful, he can step in and reap a similar return but with much lower risk. Einstein, the typical entrepreneurial optimist, continued to prod Baruch, with a little bit more vigor and aggressiveness. If he scared Baruch away, there would not be much further damage, as their opportunity was quickly fading away anyway. In the fall of 1915, another technical evaluation from Einstein showed great promise from the site. Baruch countered by sending his own expert again to evaluate the claims. In early 1916, Allen and Meyer (original partners of Einstein; Harrison had earlier sold out) sold part of their shares to Baruch at the latter's terms, and giving him an option on the remainder. The time albatross had taken their toll on them. After seven years of seeing no return on their money, they needed to recoup what they could, before it was all gone. Unlike the investors in Union Sulfur, the investor group at Big Hill did not have deep pockets to stay in the game.

Though he now controlled much of the company, Baruch was still reluctant to put up the funds himself to build the mine. He turned to J.P. Morgan & Company to finance the capital. Morgan took over 60% of the company, and the investment was made in the fall of 1916. Einstein had, at last, succeeded. By a cruel twist of fate, shortly afterwards, he died of a heart attack in November 1916, and never witnessed sulfur production from his project. Whereas, the case of Union Sulfur typified one where the investors became trapped by the entrepreneur, the situation with Texas Gulf Sulfur was the reverse. The investor trapped the entrepreneur. Baruch did very well on his reluctant investment.

Fascinating stories develop between the desperate entrepreneurs and the reluctant investors. Each venture is unique in its detail, but similar in its development. Between the exuberance of the plans of the project and the ecstasy of the financial rewards of the enterprise lies a rollercoaster ride of stress, depression and despair, intermittent with brief periods of hope and anticipation. Unfortunately, most ventures don't have the resources to hire a full-time psychiatrist to professionally assist those with the venture syndromes in this transitional period.

7.2 THE REVOLUTIONARY TECHNOLOGY:
TO SWITCH OR NOT TO SWITCH, THAT IS THE QUESTION

Leblanc had introduced a breakthrough technology that, allegedly, industry was pining for. Yet, his venture failed, and others were slow to develop in France. While market pull contributes to new technologies, they usually fail from market push. Contrary to R&D propaganda, many more technologies originate from technology push than market pull. The market is a lousy source of new ideas, because the generic response from the market on R&D development is only incremental improvements, especially to make it cheaper. Without doubt, it is nice to have inspirational market input. However, the market doesn't know what it wants until it has it! Even then the customer is often blind to what the true potential of the technology is. Technology push is an essential aspect of research, which really means teaching the customers the benefits of the (always great) invention. The influence of market pull on most research is over-rated. In many cases claiming otherwise, the so-called "market pull" only became apparent after the market finally learned what the invention could do. In other words, the market-driven aspects of many inventions are erroneously assumed by their later commercial success.

Within the sulfur world, there is a notable example of pure market pull: the replacement of elemental sulfur by pyrites after the TAC incident. An unprecedented feature was the speed of the entire process, from the technical development to its penetration into a global industry. The stage had been perfectly set. The new technology was fairly easy to develop and implement. The special feature was that the customers detested the current supplier. The sulfur world had more than its fair share of these market-pull technology advances; the Frasch process is another notable example. That is not to say that sulfur scientists are more market oriented than others. Quite the opposite, the driving force has often been the poor performance of the sulfur industry, which creates these rare market-pull opportunities. Instead of being labeled market-pull situations, fundamentally they should more properly be tagged supplier push!

The market does play a critical role in the development of new technology, but in a negative, not positive, manner. More common than market pull for technology is market push against technology. There are three major reasons for this technology barrier, cynicism, fear and laziness. Technology from a supplier is greeted by cynicism, much in the same manner that entrepreneurs face with prospective investors. A more practical concern is that there will be "bugs" in the new technology that will cause the customer downtime and lead to loss of some of their customers. The last factor is that employees of the customer will have to do extra work to be able to adapt the new technology to their process and to prove the benefits. While this can be an important factor, it is often exaggerated or simply an excuse. A prime example of market push against technology was the struggle of Muspratt when he opened his first Leblanc plant. If there ever should have been a strong case for market pull, it was this one, at least the scientific world thought so. Scientists across Europe were racing to find a solution to the high-cost, unreliable supply of alkali to support the fast growing industries of the late 18th and early 19th century. Here comes Muspratt with this new exciting and proven French technology that answers all their problems; yet, he has to give away the product to get the customers interested. He had to teach them the value of his product before they would risk changing from their safe, traditional supply. His technology push overcame the market push against change, eventually converting it into market pull.

Leblanc had invented a process and commercialized the technology, but he never came close to creating a financially-sustainable (i.e., commercially successful) business. His operations were a dismal failure, through no fault of his own. An industry, both for soda ash and also for its main raw material sulfuric acid, had been created, but the small French industry was still relatively boring. Later generations would not have taken any notice to it if it had remained such a simple domestic concern. Two critical steps had taken place, but more had to be done to reach industrial greatness.

The last, most striking development especially belonged to the British; a surprising outcome considering the exciting chemical technology had been invented and first installed outside of their country. France had a huge competitive edge, having some of the greatest chemical minds of the day and the advantage of the initial discovery. The French can not be accused of lacking business savvy, as plants using the Leblanc process were quickly built in their country. What then had left the ultimate commercial victory to the British? There may be some truth behind the suggestion that the French were more intellectual, reflected in the attitude of their greatest chemist Antoine Lavoisier (1743 - 1794), while the British were more entrepreneurial. The more rigid centralized control of education and business in France was also detrimental to more bold and risqué investments. Certainly the turmoil of the Revolution hurt progress in France, not only for Leblanc, but others in more fatal ways, such as Lavoisier. Without doubt, the British had a geological advantage with this technology as deposits of salt and coal were available in close proximity to each other. Other raw materials had to be imported, and the British merchant marine was the greatest in the world. A more influential factor, though, was strategy and business philosophy. While the French industry grew to meet its own domestic needs, the British strategy was to supply the world, the concept being spurred on by their Empire. Once economies of scale had been achieved to do this, they naturally had a cost-competitive advantage over their continental European counterparts.

The Leblanc industry of Britain was an entrepreneurial dichotomy. While their exploitation of the foreign technology is applauded as a notable example of entrepreneurship, their later abhorrence of a superior foreign technology (i.e., Solvay) is the antithesis of this same principal. Only two generations separated the two. Had the "grandsons" of the business founders of the Leblanc industry lost the entrepreneurial forte of their forefathers? Times had changed and so had the businesses. The new leaders of the Leblanc companies had a huge investment to protect. In an excellent, and somewhat desperate, strategic move, the major Leblanc companies came together in a mega-merger to form the United Alkali Company (a total of 48 works: 42 in England, 4 in Scotland, 1 in Wales, 1 in Ireland) in 1890. Included in the merger were the Leblanc assets of the pioneering firms James Muspratt & Sons and Charles Tennant & Company.

The company's first president was Charles Tennant, who had been running the legendary company Charles Tennant & Company, founded by his grandfather. This same younger Tennant was no entrepreneurial slouch. For example, he had invested in the risky first major pyrites mines in Spain, albeit two decades earlier. While the Spanish opportunity needed an entrepreneur, the Leblanc industry needed a hard-nosed business leader to protect their investments. Tennant adjusted to the circumstances.

There had been three fundamental choices open to Tennant to respond to the economically-superior Solvay technology: divest (close or sell-out), modernize (with the new technology), or run as a cash cow (and then close plants as they could no longer generate cash on a consistent basis). United Alkali was one of the first, but certainly not the last, of the emerging commodity chemical sector to face this dilemma. Their choice of the latter option became the "standard" strategy of this type of industry, much to the chagrin of ivory-tower pundits. The prudent and logical approach is for new plants to adopt the new technology (some brownfield conversions may also take place), but the bulk of the older technology plants had to do the best with what they had. As the "cash cows" are inevitably transformed into "cash pigs," the offending plant must be closed or converted. Commodity industries are not overthrown in a technological coup d'état, they just slowly fade away.

The executive of commodity industries, such as soda ash, are commonly portrayed in "case" studies for their blatant ignorance of the "facts," whose judgment has been distorted by their greater-than-though pride and self-confidence, and, of course, their blind love of their technology. One is amazed how so many incompetent people had risen to the ranks in so many technologically-antiquated global corporations. Such a derogatory portrayal has become an urban legend, more correctly a commodity fable. Arm-chair hindsight, while powerful in its knowledge of the final consequences, is ignorant of the specific circumstances surrounding the decision-making process of the times.

Business leaders must be careful not to fall victim to the "sky-is-falling" syndrome, where new-technology panic sets in. For business to react to such a situation prematurely is irresponsible. There is a wide gap between a new technology that has a competitive edge to one that will render the traditional technology obsolete. The latter rare scenario does take place (which becomes obvious only after the fact), but seldom can a strong enough case be made soon enough to convince executives and investors that the "end of the world as we know it" is upon them. The driving force for the amalgamation of the Leblanc plants into one company, United Alkali, was not out of fear of the Solvay technology; it was out of economic necessity. During the 1880's, soda ash prices crashed. Most of the Leblanc producers were losing money. They had to do something or go out of business. This economic fact melted the antagonism between the fearfully competitive companies and pulled them together under United Alkali.

Within half a century, the strength of the British Leblanc industry, its economies of scale, had become its weakness. The industry had suddenly found itself an industrial dinosaur. How could they transform themselves into Solvay mammals? The independent Leblanc companies did not have much of an opportunity to convert even if they had wanted to. Patent protection would not have allowed them to do so until after 1886, by which time Brunner Mond & Company had been already firmly established. Adoption of the technology by United Alkali would also have forced them to sacrifice their newly-founded competitive strength (their accumulated knowledge of the Leblanc process), and switch to another technology whose operating expertise lay with their competition. [One group in America did choose this route, being led by Herman Frasch. Frasch had developed his own improvements to the process. In 1881, he formed the American Chemical Company, but the venture had failed by 1887. In 1892, Frasch tried again with the Frasch Process Soda Company, which lasted until 1905.]

In addition, a full-scale conversion by United Alkali would have taken an investment unprecedented in the history of the chemical industry. How could such a massive investment been justified? Think about this from a logical viewpoint. A CEO announces at the AGM (Annual General Meeting of shareholders) that there will be a multi-million dollar write-off because of a new technology and the company is undergoing a multi-million dollar conversion over the next few years to replace their antiquated processes. By this investment, with some luck, they might be just as good as the competition. This AGM will be his last.

The United Alkali Company chose to get as much out of the existing assets as they could. They were not blind to the reality of the situation as they knew that they could not win this battle in the long-term. That does not mean that money could still not be made in the interim. They gave themselves time to phase out the technology under their own terms. Under the circumstances, they did a marvelous job.

The consolidation of the producers was a masterful stroke and brought even greater economies of scale to the industry. Less efficient plants were closed. Best practices could be implemented at each site. One must not forget the proverbial savings in administration costs. Technical efforts to improve the industry were now centralized, under the leadership of George Lunge Ferdinand Hurter (1844 - 1898) from 1890 until his death. The consolidation naturally reduced competition. They formed a sulfur cartel with Sicilian producers forming the ASSC. Some operations switched towards producing chlorine (through the Weldon process, developed in 1870) and caustic soda, which Brunner Mond could not readily do. [Unfortunately for United Alkali, a superior technology to produce these products also soon appeared from the Castner-Kellner Company. Edmund Knowles Muspratt (1843 - 1923), one of the directors of United Alkali and the son of James, dismissed the new technology.] Without doubt, the Solvay technology ironically made the Leblanc companies more efficient even without them using the new technology per se.

The highlight of the United Alkali strategy, though, was a negotiated deal with Brunner Mond. Intense competition had pushed both sides into severe financial difficulties during the latter 1880's. An agreement was reached to divide the market and fix pricing, so that both sides could make money. By definition, if United Alkali made money, Brunner Mond made more money. Alternatively, if Brunner Mond had actively targeted United Alkali, the business hand of the latter would have been forced. The minimal damage would have been from a massive price war as United Alkali would have fought to the last penny. United Alkali would have collapsed sooner, but would it have been worth the price? Brunner

Mond, though, had something much more serious to fear: what if they had forced United Alkali to the Solvay technology? The last thing they wanted was a massive competitor with the same competitive advantages. Their agreement restricted the adoption of the Solvay technology by United Alkali. Desperate times called for desperate measures. United Alkali had sold its technological soul to the Solvay devil.

By 1910, United Alkali was still the largest employer in the massive British chemical industry, even though much of the Leblanc side of the company was already gone. By World War I, the industry had largely wound down, and in 1923, half a century after the Solvay process had come to Britain, the last plant in Britain using the Leblanc process closed. However, the United Alkali Company, itself, was still a going concern. United Alkali survived, ironically, as the largest sulfuric acid producer (Chamber process) in Britain. In 1926, United Alkali joined with their Solvay adversaries, Brunner Mond, among others, to form Imperial Chemical Industries (ICI). One of the executives of the new ICI was Max Muspratt (1872 - 1934), grandson of James and son of Edmund, and the last chairman of United Alkali. The entrepreneurial legacy of the British Leblanc (and Solvay) companies exists within this global chemical company today.

7.3 THE ULTIMATE VENTURE: THE MONOPOLY GAME

The world is littered with the fossil remains of such great firms, including those using the Solvay technology. The Darwinist reality of business is the survival of the (technically and financially) fittest. Welcome to the monopoly game, where the goal is to wipe out your competition. The game is the surreptitious foundation of all business strategy, regardless of the flowery language to suggest otherwise. Different players use varying strategies cloaked in their own jargon and certainly play the game with varying degrees of skill. Companies strive to outmaneuver each other, investing to make their products better at lower cost. While these actions are always proclaimed to be market-driven, such ostentatious claims are only for the AGM and the business press. The real impetus of "market-driven" is not to satisfy the customer per se, but to use the customer to defeat the competition. Proof of this industrial game play comes from those few companies who win and become monopolies. The phrase "market-driven monopoly" is an oxymoron. The quest for a monopoly by business is really but a game since most companies have no chance of ever achieving it. Each time a major competitive advantage is gained, they may be half way there, an obviously endless trek. The game continues to be a silent engine behind the strategies of most companies. The above is a bit overstated, the degree of which is left to be determined by the reader.

The monopoly is an even higher stage of development than the global industry. Sought by many, achieved by few, this penultimate stage of industrial evolution creates a corporation of iconic reverence. For the few that begin to believe that they can win this game the business world begins to change. As they see themselves getting closer to this goal, they become more obsessed by the game. Aggressiveness against the weakening competition increases and anti-monopoly laws may be pushed to their breaking point. The quest for the monopoly can become addictive, with, at times, fatal consequences.

The sulfur industry is among the best monopoly players, and they went beyond simply playing. Monopolies have been a common feature of this industry that, some may claim, continues today. In 1947, the FTC concluded:

> Almost the entire history of the sulfur industry has been marked by cartel agreements among the world's leading producers--particularly among the two chief sources of supply, the American and the Italian companies.

In 1940, Robert Hargrove Montgomery, a professor of economics at the University of Texas, wrote **The Brimstone Game: Monopoly in Action**. The U.S. government published its own "sequel" to this book, when the FTC issued its report **The Sulfur Industry and International Cartels**, issued in 1947. The cartel activities of the sulfur industry were also under investigation in Canada. Then Justice Minister, and later Prime Minister, Louis Stephen St. Laurent (1881 - 1973) had ordered Fred Alexander McGregor (1888 - 1972), Commissioner of Canada's Combines Investigation Act, to look into various cartels, including the sulfur industry in 1943. In November 1945, St. Laurent reported to parliament that sulfur pricing had been fixed to maintain a $16 per tonne price throughout the Depression.

Monopolies are created in several ways, as exhibited within the sulfur industry. The easy monopolies are enacted by government decree. However, what is created by governments can just as easily be taken away by governments; such was the short-lived reign of TAC. Among the most daring entrepreneurs of the Sulfur Age was Aime Taix, the founder of TAC. He had little in the way of assets but managed to convince the royal government to give him exclusive rights to the marketing of sulfur from Sicily, even at risk of a diplomatic dispute with the most powerful nation in the world. Even when the government was wavering, his aggressive stance forced them to honor the contract. He only lost his monopoly after British gunboats showed up in the Bay of Naples! The little known Taix was a sulfur entrepreneur extraordaire, arguably greater than Frasch himself!

The few cases where (legal) monopolies have been earned are usually driven by technology, such as in the case of Union Sulfur. When playing the monopoly game, technology can allow you to "advance to go," and grab that elusive monopolistic position. Such pioneering and revolutionary technologies are relatively rare, but are the common pathway to obtain market dominance. The Frasch tech-nology is an exceptional example of a technology-based monopoly. Frasch and his company Union Sulfur became one of the best monopoly-game players. He not only played the game, he set the rules. The competition had little choice in the matter; else they would not be allowed to play the game at all. Frasch, himself, summarized with some eloquent words the sequence of events after the ASSC meeting in London (H. Frasch, *J. Ind. Eng. Chem.*, **4** (2), p. 139 (1912)):

> I had arranged for the sale of our sulfur in the various Euro-pean countries, and knowing the production cost to my com-petitors, I succeeded very shortly in demonstrating that Louisiana sulfur was not a swindle. I found out afterwards that the lesson had cost the Anglo-Sicilian Company 285,000 pounds sterling-but then we were friends. Their attitude changed greatly, and they decided to go out of business…

Frasch stated that he had been generous to his "friends," that is to say the competition (ASSC). He completes his summary by concluding that he had driven them out of business. What a great spin he put on the situation: he was claiming generosity after ruthlessly squashing his opponent with Machiavellian efficiency! Image how he treated his "enemies." Why did Frasch crush his weak competi-tion?…because he could and, most important of all, he knew it!

Frasch had not charged into the meeting like a bull in a china shop, although to the ASSC it may have seemed that way. His position had been carefully crafted. He had not only known his technology and business, but also those of ASSC as well. By knowing their production costs, he knew exactly what it would take to break them. It is one thing to have the best technology; it is something else to know how far you can push it to your advantage. The situation was so lopsided, there were few limits on what he could do. The convoluted structure of the Sicil-ian industry also made it impossible for them to react, even if they wanted to. On the other hand, the ASSC was isolated in its sulfur tower. With their information, or lack of it, they had dismissed the apparently wild demands of Frasch. In fair-ness to the ASSC, they had been following the sulfur venture of Union Sulfur since 1895, but even the U.S. trade journals had written them off as late as 1898! The ASSC was still astute enough to realize that there may be something to his claims. Their director Dompe was sent to investigate their operations in 1905; ASSC then came to the realization that they were in serious trouble.

The Sicilian industry had been in the enviable position that the earth's geol-ogy had conveniently planted the largest readily available deposit of sulfur in the world under the shallow earth of their island. Only a rudimentary technology was required to acquire it. A regional monopoly developed because of its unique capacity to supply the insatiable appetite that the world had developed for sulfur in the 19th century. Early in the second half of the century, another major deposit had been found deep in the ground in the bayou of Southern Louisiana, but no one, at the time, had the technology to get at it. Over a quarter century would pass before the technology was developed and the start-up of the operations of the U.S. Frasch industry began. The new process (Frasch) had lower operational costs and the product was better quality than that of the average Sicilian producer. Afterwards, the Sicilians were no longer competitive; in other words, they were no longer financially sustainable. Not only had they lost their precious monopoly, they were doomed. How quickly the mighty had fallen.

Sicily had been handed a monopoly on a sulfur platter. The situation was described by an American traveler to Sicily in 1899 (*New York Times*, p.3, April 30, 1899):

> The capitalistic cormorant has the sulfur question by its neck, and the wicked trust [i.e., monopoly] is simply gorging itself fat on the blood of the consumer.

After a century of monopolistic bliss, *le roi du sol* of Sicily had unwittingly degraded into a defenseless monarch awaiting a palace coup d'état. Frasch was more than willing to oblige. He, not Florio, became known as the "Sulfur King." Williams Haynes described the pitfalls of the Sicilian monopoly (**The Stone That Burns**, p. 77):

> For 150 years this Sicilian industry had a virtual monopoly on the world supply of brimstone... Theoretically at least, it should have controlled the market to the great profit of the mine owners and operators... Its colorful story is a vivid illustration of the extreme difficulty of maintaining a monopolistic position even under the most favorable circumstances.

The island economy had become addicted to their antiquated ways. When the Frasch industry popped up, the self-confidence built up over a century could not be easily overturned by simple facts. Their reluctance to accept the situation is understandable. How could the mighty have fallen so quickly? Only a few years before, they ruled the sulfur world. Out of nowhere, an unknown competitor shows up who, allegedly, they have little chance of successfully competing against. The claims seemed absurd, surely they must be exaggerated. Unfortunately for the Sicilians, they were true. From a practical point, their industry was over by the end of World War I, when the Sicilians ceased to be a force in the global market place. The U.S., in 1917, produced over one million tonnes of sulfur, five times that of Sicily. In later years, the gap only widened. Afterwards, the Sicilian sulfur mines were no longer an industry but a social program.

The new technology destroyed the Sicilian sulfur industry. A similar situation had happened with the Leblanc industry. The situation for the Sicilians, though, was more dire than that of the Leblanc industry with the Solvay onslaught only a decade or two before. Unlike United Alkali, the new Frasch technology was not applicable to the Sicilian situation. Frasch, himself, had tried his process in Sicily but it would not work. The one similarity between the two situations was that an agreement was negotiated with the new technology players. Whereas United Alkali could apply some competitive threats against Brunner Mond to get a good deal, the Frasch industry had nothing to fear from Sicily. All the latter could do was hope that the Frasch industry was not too hard on them. Whatever Sulexco gave them is what they got!

Overall, the reaction of the Sicilian sulfur industry to the competitive-technology threat contrasts that of United Alkali. Whereas United Alkali did almost everything in their power to fight the Solvay process, the Sicilian industry largely pleaded for charity from their government. The outside observer is stunned by their passive stance. In fairness, Santoro of COISS did what he could. From 1922 to 1925, COISS had won the lucrative contract with the prized National Sulfuric Acid Association in the United Kingdom. Besides this faint spark, there was not much fire to the Sicilian effort.

What if they had modernized? What if they had consolidated? Would the outcome have been different if…? Or, would they have simply wasted their investment in time and dollars? The "what if" scenario is so easy to apply after the fact, as arm-chair critics are protected from the subtle complexities of the situation. The Leblanc industry in Britain, for example, had modernized and consolidated, but still succumbed to a better technology. Likewise, the cost advantage of the Frasch producers was likely so insurmountable that the outcome was inevitable. We will never know for sure, though, because the Sicilian companies never even tried. Here lies the criticism and the difference between the British Leblanc and Sicilian sulfur industries. The former scratched and clawed to survive, while the Sicilian sulfur businesses laid down and pitifully awaited their destiny.

The entrepreneurial spirit abhors such lack of action. However, the Sicilian sulfur upheaval was no entrepreneurial situation. With the dysfunctional structure of the Sicilian industry, it was almost impossible to coordinate anything. So many years of their monopolistic position had created a labyrinth that even enveloped the social structure of the island itself. By the time the government finally gained overall control of the industry, the ramifications of drastic changes to the industry had far reaching consequences, far beyond simple commercial interests. If COISS, for example, had done what appeared to be the correct route from purely a commercial perspective, that is to consolidate and modernize the industry, Sicily would have been thrown into anarchy. The impoverished island economy had become addicted to the sulfur industry the way it was. Their only choice, the one followed, was again to prolong the inevitable and drag out the demise of their industry as long as possible.

While Frasch found it relatively easy to undermine the monopoly of the Sicilians, he took on the more challenging task of replacing it with one of his own. Within a few years, the Sicilians were no longer a competitive factor. The new Frasch monopoly led this masterful businessman to miss opportunities. Imagine the sulfur powerhouse that would have been created if Frasch had taken the options offered to him on Byranmound and Big Hill. However, Frasch and his company, Union Sulfur, felt impenetrable behind their technology and their massive operations. In another two decades, they were no longer in the sulfur business! The monopoly game had claimed another, one of its best players.

Even after the Frasch patents expired, a monopoly-like situation continued. A virtual monopoly often develops in oligarchic industries, such as the U.S. Frasch industry. While not a full monopoly, the industry can reap the benefits and suffer the shortfalls common to this position. Export consortiums are also virtual monopolies, especially with natural monopolies, where a rare resource base is found within a limited region, such as the case of ASSC in Sicily and Sulexco in the U.S.

Winning the monopoly game does come with its drawbacks. If the great entrepreneur Frasch could fall to its temptations, anyone could. Since the drive towards a monopoly for a company leads to improvements, monopolistic companies must stagnate by definition, because of a lack of competition. As the words imply, *competition* forces companies to be more *competitive*. The game had been driving business to improve, but it is now over. The basis for the company's business strategy has evaporated! When competition is eliminated what do you do then? There is no place to go! So few ever reach, or even truly hope to reach, a monopolistic position that the answer to this basic question is largely unknown. Most companies just enjoy it…as long as they can.

The monopoly becomes like a wasting disease that slowly takes over, as inefficiencies creep into the company that may not even be noticed by those running the firm. This decaying situation is hidden by the strong financial performance that accompanies most monopolies. High returns hide many corporate sins. Administration costs often become bloated, but this is the least of the problems. In a monopoly, there is no incentive to meet more than basic customer demands. Companies become more arrogant and less responsive to their customers, although they always claim otherwise. Investments for everything but the AGM and annual report decline. Both capital and R&D suffer. The former results in an aging production base and the latter increases the likelihood that the next generation technology will be invented by someone else.

This process is self-correcting, as the monopoly becomes more vulnerable. The ineptitude and arrogance of the rotting firm encourages others to overthrow it. Customers are mainly passive during this transition, waiting for something better to come along. When that happens, they will enthusiastically embrace the new entry. In the worst case of a monopoly abusing, in the eyes of the market place, their exclusive position, the customers, themselves, must take action. They will actively search for, or even develop themselves if necessary, a new supplier. A famous example is the replacement of elemental sulfur by pyrites, as discussed above.

Among sulfur aficionados, Frasch is still looked upon with an almost divine reverence. From his basic patent of 1890, evolved more than a global industry; he had won the elusive monopoly. The inventor, himself, was directly responsible for its creation. Frasch himself was called the "Sulfur King," a fitting epithet for the founder of the modern industry of *le roi du sol*. Frasch was the great intellectual that brought in a sulfur renaissance, which transformed a stodgy industry into an enlightened, world-class enterprise. The Sicilian industry was barbaric by comparison. This empire was later ruled by a business triumvirate, Union Sulfur, Freeport Sulfur and Texas Gulf Sulfur. After the premature demise of the former, the latter two were left to rule a global industry for almost half a century, and they did a marvelous job. Exceptional leaders of the time were Swenson and his deposer Williams, and their competitor Aldridge. The vast empire was kept under control by the effective management of Sulexco. The only regions that they did not conquer were ones that they did not want. The world was theirs! The yellow empire of the monopoly of Frasch America was looked upon with envy, spite, fear and awe. The performance of these companies defied the Depression and global wars. Returns were so high that they were (almost) sinful. In a commodity industry has there ever been such dominance as had been achieved by these lone two U.S. sulfur companies?

7.4 WHO KILLED THE SULFUR ENTREPRENEUR?

Even this industry inevitably succumbed to economic fate. As the sulfur possessions of Sicily had once fallen to the economic competitive onslaught of the Frasch process, so too the Frasch industry fell to the economic competitive hordes of recovered sulfur. They first came in the thousands of tonnes, then by the millions. Texas Gulf Sulfur and Freeport Sulfur valiantly fought to the bitter end. When Main Pass closed in 2000, Frasch had finally fallen after a glorious century.

To the nostalgic sulfur purist, the industry has now fallen into a dark age. While such bitter feelings are exaggerated, the golden glow of sulfur has indeed faded. This statement is not a criticism of the oil and gas industry, but it does reflect the reality of the situation. Sulfur and oil (and gas) don't mix. Their connection is not by desire but by necessity. When sulfur prices and demand are high, sulfur may be promoted to the position of a co-product. Usually, though, it is a by-product to the oil and gas producers, at the best of times. When demand is poor and sulfur is poured into block, *le roi du sol* is sadly delegated to the status of waste product. More times than not, it is in the latter two categories that sulfur routinely finds itself. With the demise of the Frasch industry, the sulfur crown has been tarnished. Now, (recovered) sulfur no longer enters the Board Room or the AGM. If mentioned at all in the Annual Report, a footnote appears in the smallest font. News about sulfur only intrudes into executive meetings when there is a problem, a problem that affects gas or oil. Silence is golden in today's lack luster sulfur world. Sulfur, the old warhorse of chemistry, has been sent to the glue factory, as it is relegated to a cost of doing business. What has happened to Xanthus!

While there are a few notable exceptions, most oil and gas companies consider the "Sulfur Department" to be organization-chart purgatory. An official proclamation bellows from the offices somewhere up in the stratosphere that this is not their core business! Such demeaning announcements relegate those in these departments to the lowest levels of the business caste system. When selling a global commodity with low selling prices into a highly competitive marketplace, the actual skills and creativity to make the business a success are far greater than with "normal" businesses. More so for those in the sulfur industry today, where employees have to face extra daunting challenges. A typical sulfur "marketing" department is a person with a phone and a computer. The same person will likely be the Business Manager and bottle washer. When a product sells for only a penny a pound, and must be shipped around the world, the freight bill outweighs the value of the product. Anyone can make huge profits when oil is $75 a barrel, but how does one make money on a product that does not sell for this much per tonne? At times, this is the delivered cost to China! Often isolated from most of the financial support and intellectual resources of the corporation, with limited budgets and staff, only the good survive and the best thrive in the desolate confines of the Sulfur Department.

The industry, though, must be diligent. Beware of the yellow element, for it is an albatross around the neck of the oil and gas industry, especially in Athabasca. What will be done with the corresponding massive production of sulfur? The situation is a double-edged sword, which produces a conflict of interest. The sulfur burden is a threat to future oil sands development. A cost effective measure to remove this material must be developed. At the same time, the release of these massive volumes into the sulfur marketplace is a threat to current recovered sulfur supplies. The conflict of interest arises from the fact that most of the companies producing and marketing recovered sulfur today from sour natural gas are also operating in the oil sands.

Something has been lost in the new sulfur world. Who killed the sulfur entrepreneur? This question is actually a warning to the oil and gas industry. A Sicilian-type complacency has taken over. The modern industry seeks, no demands, status quo. Such corporate indifference to sulfur is a threat to their core industry. What will the industry do when environmental regulation forces the release of huge volumes of sulfur stockpiled at the oil sands of Northern Alberta? Will an entrepreneur short-circuit this process and find a way of bringing the sulfur economically into the global marketplace? The sulfur world, especially recovered sulfur, has not seen much entrepreneurial spirit for the past few decades. Such dormant times are only a transition before a new generation of entrepreneurs sweep into this industry. Who will be the 21st century Frasch to introduce a new renaissance in the sulfur world?

REFERENCES

GENERAL

Haynes, W.: **The Stone That Burns**, *D. Van Nostrand Company*, Norword, Mass., 1942.

Haynes, W.: **Brimstone, The Stone That Burns**, *D. Van Nostrand Company*, Princeton, NJ, 1959.

Haynes, W.: **American Chemical Industry, A History**, Volume II, 1912 to 1922, *D. Van Nostrand Company*, New York, 1945.

Haynes, W.: **American Chemical Industry, A History**, Volume VI, The Chemical Companies, *D. Van Nostrand Company*, New York, 1949.

Ober, J.: **Materials Flow of Sulfur**, U. S. Department of the Interior, U.S. Geological Survey,

Various, **Canadian Minerals Yearbook**, Natural Resources Canada, 1944 to 2003; available on-line: http://www.nrcan.gc.ca/mms/cmy/com_e.html#Stone

Various: **Minerals Yearbook**, U.S. Geological Survey (U.S. Bureau of Mines), 1932 to 2004; available on-line: http://minerals.usgs.gov/minerals/pubs/commodity/sulfur/

EARLY HISTORY

Agricola, G. (Brandy, M.C., and Brandy, J.A., trans.): **De Natura Fossilium**, *Dover Publications*, Mineola, NY, 2004.

Agricola, G. (trans. Hoover, H.C., and Hoover, L.H.): **De Re Metallica**, *Dover Publications*, Mineola, NY, 1950.

Anon: **Annual Report of the Federal Trade Commission for the Fiscal Year Ended June 30, 1947**, United States Government Printing Office, Washington, 1947.

Biringuccio, V. (trans. Smith, C.S. and Gnudi, M.T.): **The Pirotechnia**, *Dover Books*, 1990.

Cardwell, D.: **The Fontana History of Technology**, *Fontana Press*, London, 1994.

Dickinson, H.W.: *Transactions of the Newcomen Society,* **18**, p. 43 to 60, 1937/1938.

Grant, James: **Bernard M. Baruch, The Adventures of a Wall Street Legend**, *John Wiley & Sons*, Toronto, 1997.

Hou, T-P.: **Manufacture of Soda,** *The Chemical Catalogue Company*, New York, 1933.

Lundy, W.T.: *J. Ind. Eng. Chem.*, **42** (11), p. 2199, 1950.

Mellor, J.W.: **Comprehensive Treatise on Inorganic and Theoretical Chemistry**, Volume X, *Longmans, Green and Co.*, New York, 1930.

Pliny (Rackman, H., trans.): **Natural History**, IX, *Harvard University Press*, Cambridge, 1952.

Trevelyan, R.: **Princes Under the Volcano,** *Phoenix Press*, New York, 1972.

Warrington, C.J.S., and Nicholls, R.V.V.: **A History of Chemistry in Canada**, *Sir Isaac Pitman and Sons Ltd.*, Toronto, 1949.

HERMAN FRASCH

Chandler, C.F., *J. Ind. Eng. Chem.*, **4** (2), p. 132, 1912.

Frasch, H: J. *Ind. Eng. Chem.*, **4** (2), p. 134, 1912.

Haynes, see above.

Lucas, A.F., *J. Ind. Eng. Chem.*, **4** (2), p. 140, 1912.

Pough, F.H., *J. Ind. Eng. Chem.*, **4** (2), p. 143, 1912.

Sutton, W.R.: **Herman Frasch**, Dissertation, *Louisiana State University*, May 1984.

SULEXCO AGREEMENTS

UVZI – 1934

AGREEMENT

BETWEEN THE SULFUR EXPORT CORPORATION AND THE UFFICIO PER LA VENDITA DELLO ZOLFO ITALIANO
Dated August 1st, 1934

This agreement made this first day of August 1934, between the Sulfur Export Corporation of the State of Delaware, United States of America, hereinafter called the Export Corporation party of the first part, and the Ufficio Per La Vendita Dello Zolfo Italiano a compulsory sales office for all Italian crude sulfur, created by Italian Royal Decree Law of 11th December 1933, No. 1699, hereinafter called the Ufficio party of the second part, withnesseth:

Duration.---This agreement shall continue from year to year beginning August 1st, 1934, it being understood however, that either party may terminate this agreement at any date upon giving six months notice in writing to the other of such intention.

It is also understood and agreed that in the event of any material increase in the present production of elementary sulfur from pyrites this agreement shall be cancelled and become null and void three months after commencement of such increased production.

It is also understood and agreed that in the event of any new source or new production of elementary sulfur other than from pyrites as heretofore provided arising outside the control of either party which in the opinion of either party may be considered as nullifying or materially detracting from the benefits derived from this agreement then the matter shall be promptly discussed and if no agreement pertaining thereto is reached this agreement shall automatically expire at the end of the third month thereafter.

It is also understood and agreed that in the event of a material change occurring in the present gold value of the U.S. dollar which in the opinion of either party may be considered as nullifying or materially detracting from the benefits derived from this agreement then the matter shall be promptly discussed and if no

agreement pertaining thereto is reached this agreement shall automatically terminate thirty days after such failure to reach agreement.

Tonnage.---In all the following articles of this agreement dealing with the
division or allocation of tonnage, all exports made by other Italian producers or
sellers of sulfur shall be included as part of the quota of the Ufficio; and the Uffico
guarantees that Italian producers and sellers of sulfur other than the Ufficio will
not export sulfur for acid-making at a reduced price. All exports to the markets
covered by this agreement of sulfur produced by the Jefferson Lake Oil Company
from their sulfur mine in Iberia Parish in the State of Louisiana and of sulfur produced by the Duval Texas Sulfur Company from their Palangana mine in the State
of Texas shall for the purpose only of determining quotas and tonnage be considered as made by the Export Corporations.

Markets.---This agreement refers to and covers all sales of sulfur made by the
two parties to all countries of the world excepting only:
1. The Kingdom of Italy, its dependencies and colonies and
2. North America, Cuba, the Islands off the coast of Canada and the insular pos-
    sessions of the United States of America.

Division of tonnage.---Subject to the conditions and agreements herein set
forth all export sales of the two parties to the markets covered by this agreement
shall be apportioned on the basis of allowing---
1. The Export Corporation and the Ufficio each 50% of the first 480,000 tonnes
    of annual invoiced sales;
2. The Export Corporation 75% plus 5,000 tonnes and the Ufficio 25% minus
    5,000 tonnes of annual invoiced sales in excess of 480,000 tonnes up to
    625,000 tonnes;
3. The Export Corporation 90% and the Ufficio 10% of annual invoiced sales in
    excess of 625,000 tonnes.

The expression "annual invoice sales" in this clause shall be understood to
include all exports of manufactured sulfur from the United States and from the
Kingdom of Italy to all countries covered by the agreement as above. For the purpose only of determining quotas or tonnage as herein set forth all exports of manufactured sulfur from the United States and from the Kingdom of Italy to all
countries covered by the agreement as above shall also be included.

In making current determinations of allocation of crude sulfur the exports of
manufactured sulfur from the Kingdom of Italy shall first be deducted from the
share of the Ufficio and similarly the exports of manufactured sulfur from the
United States shall first be deducted from the share of the Export Corporation.

For example, assuming the total amount of invoiced sales of crude sulfur plus
exports of manufactured sulfur for one year under this agreement is 480,000
tonnes of which amount 70,000 tonnes is manufactured sulfur exported from the
Kingdom of Italy and 5,000 tonnes is manufactured sulfur exported from the
United States the proportion of each of the parties of the total 480,000 tonnes

being 50% or 240,000 tonnes, there shall be deducted from the Ufficio's quota of 240,000 tonnes the amount of manufactured sulfur exported from the Kingdom of Italy, to wit 70,000 tonnes, leaving a balance of 170,000 tonnes of crude sulfur to be sold by the Ufficio; there shall likewise be deducted from the Export Corporation's quota of 240,000 tonnes the amount of manufactured sulfur exported from the United States to wit 5,000 tonnes leaving a balance of 235,000 tonnes of crude sulfur to be sold by the Export Corporation making a total of 405,000 tonnes of crude sulfur to be sold by the parties jointly, the percentages being approximately 41.97% for the Ufficio and 58.03% for the Export Corporation.

For second example, assuming the total invoiced sales for one year to be in excess of 480,000 tonnes but less than 625,000 tonnes, say 580,000 tonnes, 50% of the first 480,000 tonnes gives the Ufficio 240,000 tonnes plus 25% less 5,000 tonnes of remaining 100,000 tonnes, to wit 20,000 tonnes, a total Ufficio quota of 260,000 tonnes, and assuming the same figures for manufactured sulfur as in the preceding example, then the amount of crude sulfur to be sold by the Ufficio would be 190,000 tonnes and the amount of crude sulfur to be sold by the Export Corporation would be 315,000 tonnes, the percentages being approximately 37.62% for the Ufficio and 62.38% for the Export Corporation.

For the third example, assuming the total invoiced sales for one year to be in excess of 625,000 tonnes, say 725,000 tonnes, 50% of the first 480,000 tonnes gives the Ufficio 240,000 tonnes, 25% minus 5,000 tonnes of 145,000 tonnes gives the Ufficio 31,250 tonnes, and 10% of 100,000 tonnes gives the Ufficio 10,000 tonnes, being a total Ufficio quota of 281,250 tonnes, and assuming the same figures for manufactured sulfur as in the two preceding examples then the amount of crude sulfur to be sold by the Ufficio would be 211,250 tonnes and the amount of crude sulfur to be sold by the Export Corporation would be 438,750 tonnes the percentages being approximately 32.50% for the Ufficio and 67.50% for the Export Corporation.

Sulfur for the manufacture of sulfuric acid.–It is the judgement of both parties that the sale of a certain tonnage of sulfur at a special price solely for the manufacture of sulfuric acid is in their mutual interest. Any such sales of sulfur for the manufacture of sulfuric acid shall be made only by mutual agreement of the parties and the terms and conditions thereof shall likewise be mutually agreed. The Export Corporation having made known to the Ufficio the existence of a contract dated February 1st, 1934, between the Export Corporation and the National Sulfuric Acid Association, Ltd., London, covering a sale of 100,000 tonnes for delivery over two or 2½ years at the said Association's option, the Ufficio undertakes during the life of this agreement to supply a proportion of the sulfur to be delivered against the said contract at the price and under the terms therein stated and other details and conditions as to sampling and analysis to be agreed upon, the said proportion of the Ufficio to be:

1. If the invoiced sales under said contract do not exceed 35,000 tonnes in any one year the Ufficio will furnish 50%;
2. If such invoiced sales exceed 35,000 tonnes in any one year the Ufficio shall supply its share of such excess over 35,000 tonnes in the proportion laid down in section (2) of the article of this agreement entitled Division of Tonnage.

The tonnage supplied by both parties under the said contract shall from part of the total annual invoiced sales as figured in the article hereof entitled Division of Tonnage.

Effective date.–The effective date of this agreement is August 1st, 1934. The first year under this agreement shall begin on such date and shall end on July 31st, first year under this agreement shall begin on such date and shall end on July 21st, 1935; all succeeding years shall begin on August 1st and end on July 31st. All shipment and/or deliveries made on and after August 1st, 1934, either from stock in warehouse or otherwise shall be included in the computation of tonnage and quotas.

Prices.–The prices, which at all times are to be such as will foster the sale of high-grade sulfur produced by the parties hereto in competition with pyrites and/or sulfur produced by others and/or substitutes for elementary sulfur, shall, together with terms and conditions of all sulfur sold under this agreement, be fixed from time to time by the parties having regard to changing conditions, in such manner as best to serve their mutual interest.

Allocations.–The allocation or distribution of tonnage sold under this agreement shall be fixed from time to time in such manner as may afford each party the advantages of freight rates and market conditions arising by reason of geographical location with regard to the market served, insofar as this may be done without prejudice to the other party; but each party shall be entitled to its proportionate share of crude sulfur under this agreement of any market upon request to the other, and each party shall be obligated to take its said proportionate share in any low-priced markets upon request of the other.

In order to effectuate this purpose and to facilitate the operations to the mutual and best interests of both parties, they shall each appoint an assistant executive representative, resident in Europe who shall constitute a central bureau for the exchange of data and compilation of statistics and shall assist in the allocation and distribution of tonnage and have such other functions as shall from time to time be assigned them by the parties for the furtherance of the purposes of this agreement. Each party shall promptly furnish to the Central Bureau copies of all contracts and invoices made by it and a note of shipments to warehouses and shipments of manufactured sulfur in the business covered by this agreement. Where in competitive markets the actions of the agents of the respective parties may become detrimental to the best interests of either of the parties hereto all the business under this agreement shall be transacted through the Central Bureau for such time as may by the parties be decided upon as proper.

Adjustments of tonnage and of sales shall be made for the period ending January 31st, 1935, and or each successive period of six months. If at the termination of any such six-month period it is determined that either party has not sold its full quota as provided herein, or that adjustments in distribution and allocation of tonnage are necessary, such deficiency and necessary adjustments shall be made up within the next succeeding six-month period.

If at any time this agreement shall come to an end the final adjustment shall be made in cash at the average price f. o. b. the respective shipping ports for the period to which the adjustment refers, realised under this contract by the party to whom the cash shall be paid. The party to whom the cash shall be paid will hold the corresponding tonnage of sulfur at the disposal of the other party for six months free of any charge including insurance, and will deliver it f. o. b. the respective shipping ports at the other party's request. This tonnage of sulfur shall not be sold in the territories excluded from the world's crude sulfur market as defined under this agreement.

Statistical.–On or before the thirtieth of each month each party shall furnish to the other, and to the Central Bureau a statement covering the operations of the preceding calendar month which shall show the total tonnage shipped, total tonnages sold and total tonnage delivered, destinations, prices realised, both f. a. s. and/or f. o. b. and/or c. i. f., and/or c. f., freight rates paid and such other information as may be from time to time necessary for proper forecast and allocation.

Penalty for violations.–In case either party shall directly or indirectly export any sulfur or permit the export of any sulfur to the territories covered by this agreement otherwise than as herein provided, for each tonne so exported there shall be a reduction in such offending party's allocation provided for herein of two tonnes, and an increase in the allotment of the other party of two tonnes.

Manufactured sulfur.–It is the judgement of both parties that the situation of the sulfur-manufacturing industry in the countries covered by this agreement should be maintained as it at present exists throughout the life of this agreement; each party agrees not to do or encourage anything which would result in altering such present situation and any action of a nature to alter such present situation shall be jointly considered and both parties shall use their best endeavours to present any such alterations.

Force majeure.–If by reason of force majeure either party is unable to ship its yearly quota or tonnage in such event a revision of allocation to meet the situation as created thereby will be made so as to adjust the matter equitably.

Arbitration.–In case of a disagreement arising out of any matter in connection with the construction of this agreement or performance thereof, which it may be found impossible to settle by amicable arrangement, the same shall be submitted to a Board of Arbitration to sit in London consisting of three members; one chosen by the Export Corporation, one by the Ufficio, and an Umpire to be chosen by the joint agreement of the first two arbitrators.

In case of any party failing to appoint its arbitrator within fifteen days of the notice of the other party so to do, the party who has appointed an arbitrator may appoint that arbitrator to act as sole arbitrator in he reference, and his award shall be binding on both parties as if he had been appointed sole arbitrator by consent.

In case of the first two arbitrators failing to agree within fifteen days of their nomination as to the appointment of the Umpire, such appointment shall be made by the President for the time being of the London Chamber of Commerce upon request of either of the parties.

Insofar as not specially provided for in the present agreement the arbitration shall be subject to the English Arbitration Act of 1889 or any statutory modification thereof at the time subsisting.

Any award shall be final and binding upon both parties.

Legalities.–The Agreement shall not be, nor be construed to be in any respect as a partnership or agreement between the Corporations holding shares of stock in the Export Corporation and the Ufficio nor to bind any of such Corporations in any way individually nor the Italian Government.

It is agreed that this contract shall be deemed to have been executed and delivered at the City of London, England, and the interpretation and enforcement thereof shall be governed by the provisions of the English law as it shall from time to time exist and it is agreed that, subject to the provisions hereof respecting arbitration, jurisdiction shall be given to the English Courts to take cognisance of disputes hereunder and to render judgement or decree which shall be binding upon the parties.

Notice and service of process.–Any notice provided to be given hereunder may be given in writing either by delivering the same in a sealed envelope addressed to the Central Bureau representative of the party to be served at the office of the Bureau, or by Service upon such representative in the manner provided by the laws of England for service of legal papers.

In order to make effective the provisions hereof in respect of arbitration and procedure in the English Courts each party shall at all times maintain in the City of London a person or corporation upon whom legal process may be served on behalf of the other party in the manner provided by the laws of England for the service of legal papers with the same force and effect as if due service had been made upon either party at its home office in the country of its incorporation or institution.

In witness whereof both parties have hereunto set their hands and seal the day and year first above written, the present being one of three triplicate originals so executed.

SUFLFUR EXPORT CORPORATION,
BY C.A. SNIDER, President.
Attest:
B.C. HUGHES.

UFFICIO PER LA VENDITA DELLO ZOLFO ITALIANO,
BY C. ANGELLI, President.
Attest:
VESPUCCIO CIUCCI.

AUGUST 1ST, 1934.

From: Sulfur Export Corporation, New York.

To: The Ufficio per la Vendita dello Zolfo Italiano, Rome.

GENTLEMEN: Referring to the agreement this day entered into between us, we herewith set forth certain additional and supplementary matters which have been agreed upon between us, as follows:

(1) *Duration.*–In connection with the provision for cancellation in the event of any material increase in the present production of elementary sulfur from pyrites, specifically such present production is:

	tonnes per annum
Orkla	70,000
Rio Tinto	30,000
Mason & Barry	10,000
	110,000

(2) *Sulfur for the manufacture of sulfuric acid.*–For the purpose of facilitating the agreed division of tonnage to be supplied to the National Sulfuric Acid Association, Ltd., under the contract cited in the article headed "Sulfur for the Manufacture of Sulfuric Acid," the Central Bureau will from time to time receive from the National Sulfuric Association, Ltd., requests for shipments and will then allocate such tonnage to the parties in such manner as is most convenient to the parties and to the National Sulfuric Acid Association, Ltd.

Such allocations to the Ufficio will be evidenced by purchase requisitions signed by B.C. Hughes on behalf of the Sulfur Export Corporation as per sample form attached.

(3) In connection with the article entitled "Prices," the following schedule will be effective until changed by mutual agreement of the parties:

Except in special cases which may be presented and agreed upon all sales to be made c. i. f. The prices shall be in United States dollar currency or equivalent, cash payment---

C. i. f. ports in Europe, except ports in Lithuania, Poland, the port of Danzig, ports in Portugal, France, Spain, Belgium, Yugoslavia, Albania, Romania, Greece, Turkey in Europe, Bulgaria, and U.S.S.R.	$23.00
C. C.i. f. ports in France and Belgium	23.50
C. i. f. ports in Lithuania, Poland, the port of Danzig, ports in Portugal, Spain, Yugoslavia, Albania, Romania, Greece, Turkey in Europe, Bulgaria, and U.S.S.R	24.00
C. i. f. ports in Asia	24.00
C. i. f. ports in Australasia	23.50
C. i. f. ports in South America	23.00
C. i. f. ports in Africa excluding Algeria	24.00
C. i. f. ports in Algeria	23.50
C. i. f. any other ports not covered above	23.50

Such prices are understood to be minimum net per tonne of 2,240 lbs., delivered weight, and refer to the highest Italian grade (Gialla Superiore---Best Yellow) and to the American quality of crude sulfur.

All sales of lower grades of Italian sulfur are to be made according to the above schedule, conditions and terms, except that a discount or differential may be allowed from the schedule in no case exceeding: $0.85 per tonne for Gialla Inferior (Inferior Yellow); $1.55 per tonne for Buona (Good) quality; $2.25 per tonne for Corrente (Current) quality.

(4) In connection with the article entitled "Allocation," the Central Bureau shall be located in London, England, and all expenses thereof except such salary or remuneration as may be paid to the representative thereon of the Ufficio shall be borne by the Sulfur Export Corporation.

To assist in the proper allocation of tonnage each party will promptly file with the Central Bureau a schedule of unfilled orders and copies of all uncompleted contracts as of August 1st, 1934 and copies of all engagements to supply sulfur on or after August 1st, 1934.

In connection with the article entitled "Notice and Service of Process" the Sulfur Export Corporation has conferred upon Mr. Bertie Cameron Hughes of London House, 35 Crutched Friars, London, E.C. 3., authority to accept service of legal process for it and hereby designates him as the party upon whom legal process under the agreement may be served to bind it; and the Ufficio hereby gives authority to Henry Gardner & Co. Ltd., of 2 Metal Exchange Buildings, London, E. C. 3., to accept legal service for it in England and hereby designates said Corporation as the party upon whom legal process under the agreement may be served to bind it.

We will thank you to confirm and accept the above.

Yours truly,

SULFUR EXPORT CORPORATION,

By (Signed) C.A. SNIDER, President.

Confirmed and accepted:

UFFICIO PER LA VENDITA DELLO ZOLFO ITALIANO,

By (Signed) C. ANGELLI, President.

ORKLA – 1937

MEMORANDUM

OF AGREEMENT MADE IN LONDON THIS FIRST DAY OF APRIL 1936 BETWEEN ORKLA GRUBE A.B. OF LOKKENVERK, NORWAY, AND THE SULFUR EXPORT CORPORATION OF NEW YORK, U.S.A.

This Agreement is to take effect as of January 1st. 1937, and shall remain in force until December 31st, 1941, unless sooner terminated as hereinafter provided.

This Agreement is confined to the Continents of Europe, Asia and Africa including adjacent islands, hereinafter called the joint territory.

It is agreed that sales in the joint territory made by the Sulfur Export Corporation and the Orkla Company shall be divided one-third to the Orkla Company and two-thirds to the Sulfur Export Corporation.

Commencing January 1st, 1937, sales will be proportioned in this ratio as nearly as may be and should either of the parties be behind in its tonnage that tonnage will be made up during the next year.

Information will be currently furnished by both parties to the other advising all sales which each party had made in the joint territory. The price at which these sales shall be made shall be mutually agreed upon at a meeting, which shall take place in London not less than three months before expiration of each calendar year. If at this meeting no agreement is reached between the two parties as to prices for the coming year then either party may elect to cancel this Agreement to take effect on the following December 31st. The party desiring cancellation shall give notice to the other in writing to be mailed prior to November 1st.

It is contemplated that as soon as Orkla increases its production of brimstone above 70,000 tonnes per year and thereby releases to the pyrites industry any tonnage of pyrites which Orkla is now selling an endeavour will be made as soon as feasible through collaboration with the pyrites industry as soon as such tonnage is released to secure an equivalent brimstone tonnage in the joint territory. In no case without agreement between the parties shall the price of brimstone be less than 75/ ---c. i. f. and in no case shall an attempt be made to secure customers who would be in direct competition with present users of brimstone without the mutual consent of both parties.

Orkla agrees as far as lies in its power that it will not license to others the use of the Orkla process covered by patents which are now controlled by the Aktiebolaget Industrimetoder, Stockholm, nor to assist others in the development and use of the process other than as already committed.

It is recognised in principle by both parties that it may be desirable to secure through purchase or otherwise certain patents in the joint territory, which may have a potential value in the production of sulfur in the joint territory.

If any such patents are acquired they shall be acquired jointly by the two parties, payment being two-thirds by the Sulfur Export Corporation and one-third by Orkla. The title to such patents shall be held in trust by A. B. Industrimetoder, a subsidiary of Orkla which Company will endeavour to protect these patents for the mutual benefit of both parties and at their mutual expense in the proportions mentioned. A. B. Industimetoder shall issue no license for such jointly owned patents without the consent of both parties.

It is understood that as far as practicable sales of Orkla Sulfur shall be confined to Scandinavia and the countries bordering on the Baltic.

In case other American producers outside the Sulfur Export Corporation shall ship sulfur into the joint territory it shall be the option of the Orkla Company in case it does not wish to absorb its one-third portion of such loss of tonnage to cancel this Agreement giving to the Sulfur Export Corporation 90 days' notice of its wish to cancel.

The Orkla Company has been advised of the existence of an agreement between the Ufficio per la Vendita dello Zolfo Italiano and the Sulfur Export Corporation. This present Agreement is subject to the concurrence of the Ufficio.

In case of a disagreement arising out of any matter in connection with the construction of this Agreement or performance thereof, which it may be found impossible to settle by amicable arrangement, the same shall be submitted to a Board of Arbitration to sit in London consisting of three members; one chosen by the Sulfur Export Corporation, one by the Orkla Company, and an Umpire to be chosen by the joint agreement of the first two arbitrators. In case of any party failing to appoint its arbitrator within fifteen days of the notice of the other party so to do, the party who has appointed an arbitrator may appoint that arbitrator to act as sole arbitrator in the reference, and his award shall be binding on both parties as if he had been appointed sole arbitrator by consent. In case of the first two arbitrators failing to agree within fifteen days of their nomination as to the appointment of the Umpire, such appointment shall be made by the President for the time being of the London Chamber of Commerce upon request of either of the parties.

In so far as not specially provided for in the present Agreement the arbitration shall be subject to the English Arbitration Act of 1889 or any statutory modification thereof at the time subsisting.

Any award shall be final and binding upon both parties.

SULFUR EXPORT CORPORATION,

(Signed) By WILBER JUDSON, Vice President.

Attest:

B.C. HUGHES. (signed)

ORKLA GRUBE A. B.,

(Signed) By MARC WALLENBERG.

Attest:

N.E. LENANDER. (signed)

LONDON, April 1st, 1936.

MESSRS. ORKLA GRUBE A. B.,

Lokkenverk, Norway.

GENTLEMEN: With reference to the Agreement made between us today, it has been further agreed as follows:

1. The Sulfur Export Corporation will supply those customers substituting brimstone for pyrites as set forth in the Agreement, with, however, the right on the part of the Sulfur Export Corporation to have Orkla supply one-third of such total tonnage should the Sulfur Export Corporation so desire.

2. In the event of the dissolution of the Sulfur Export Corporation the Agreement is to terminate and become null and void as from the date of such dissolution.

"* * * in no case shall an attempt be made to secure customers who would be in direct competition with the present users of brimstone without the mutual consent of both parties."........................

It is desired to place on record that this arises from a desire to avoid the difficulties which might come about through two Works in the same line of business and/or in close proximity obtaining supplies of brimstone at widely differing prices. It is considered to be in our best interests to avoid difficulties of this nature and the object of the clause in question is to ensure that there shall be full discussion and mutual agreement before any action is taken which might be detrimental to the interests of us both.

Your signature on the duplicate hereof will constitute your acceptance of the foregoing.

Yours truly,
SULFUR EXPORT CORPORATION,
(Signed) By WILBER JUDSON, Vice President.
Confirmed and accepted:
ORKLA GRUBE A.B.,
(Signed) By N.E. LENANDER.

Source: University of Leiden, http://www.geschiedenis.leidenuniv.nl/index.php3?m=&c=443

FTC Review of Sulexco — 1947

THE SULFUR INDUSTRY AND INTERNATIONAL CARTELS

The sulfur industry is even more concentrated than the copper industry, and, like the latter, has been characterized by cartel agreements, rigid prices, and comparatively high profit ratios for the leading companies, according to the Commission's Report on the Sulfur Industry and International Cartels, transmitted to Congress June 16, 1947.

22 ANNUAL REPORT OF THE FEDERAL TRADE COMMISSION, 1947

Native or natural sulfur is mined in the United States, Italy and Japan, but as of 1946 the United States was by far the largest producer.

Texas Gulf Sulfur Co. ranks as the world's largest producer of native sulfur. During the 25-year period, 1919-43, this company produced 28,124,372 long tonnes of sulfur, an annual average of 1,124,975 tonnes. Its production was equivalent to approximately 56 percent of the total United States production during that time and, together with that of Freeport Sulfur Co., accounted for approximately 86 percent of the total United States production during the past 25 years.

Freeport Sulfur Co. is the world's second largest producer of native sulfur. During the 25 years ending in 1943, the company produced 15,030,374 long tonnes of sulfur, an average of 601,215 tonnes a year. The annual production of 1,027,837 tonnes in 1943 represented approximately 40 percent of the total United States production. For the 25-year period, the company's production was approximately 30 percent of the total United States production.

As in the case of most highly concentrated products, prices of sulfur have remained unchanged over long periods of time. The greatest variations in the price of crude sulfur occurred during and immediately after World War I, when the sales realization of marketed production, f.o.b. mines, ranged from a maximum of $22 per long tonne in 1918, to a minimum of $15.12 per long tonne in 1919. By 1926 the sales realization of marketed production f.o.b. mines became stabilized at $18 per tonne and remained at approximately that amount for the next 11 years.

The cost-price relationships in the industry have been such that the operations of domestic sulfur producers from the beginning of their operations have been highly profitable. Texas Gulf Sulfur Co. earnings during the 28-year period of its operations were equivalent to 21.97 percent of its invested capital after providing for the payment of Federal income and excess profits taxes, and 26.10 percent of the investment before providing for such taxes. Profits, after taxes, were large in all years, ranging from maximums of 62.9 percent of investment in 1926, 67.9 percent in 1927, 66.5 percent in 1928, and 58.7 percent in 1929, to a minimum of 11.7 percent in 1938. Freeport Sulfur Co. earned an average profit on its total invested and borrowed capital during the 28-year period, 1919-1946, of 13.64 percent, after providing for the payment of Federal income and excess profits taxes, and 16.35 percent before providing for such taxes. The company earned profits in all years but three, when losses were sustained equivalent to 5.82 percent on the investment in 1921, 2.53 percent in 1922, and 2.65 percent in 1924. Substantial profits were earned in most of the other years and were largest during the years 1927-1933 when an average of 24.76 percent on the investment was earned, after providing for Federal taxes. Profits were lower thereafter, averaging 13.09 percent on the investment during the years 1934-1946. During this period the range in profits was from a minimum of 10.60 percent to a maximum of 17.30 percent.

Almost the entire history of the sulfur industry has been marked by cartel agreements among the world's leading producers--particularly among the two chief sources of supply, the American and the Italian companies. The factor limiting the ability of the cartel participants to enhance and maintain prices in foreign markets was always the uncontrolled production of non-member companies. Since the early 1920's this competition has consisted of uncontrolled producers of natural sulfur, mainly by American, Japanese and South American producers, and producers of by-product sulfur in the smelting of pyrites for their metal contents, mainly in Scandinavia, Spain and Portugal under patents controlled by Orkla Grube Aktebolaget Industrimetoder of Stockholm, and in Italy by the Montecatini interests.

When competition arose from these three sources in the early 1930's, each became the object of cartel control to maintain cartel price structures and distribution arrangements. Orkla competition was controlled by a secondary agreement between it and Sulfur Export Corp dividing the foreign market, fixing foreign prices, limiting further issuance by Orkla of licenses to third parties in Europe, and providing for joint acquisition of basic patents under which other by-product competition might arise.

Montecatini (Italian) competition was disposed of by the Italian Government, compelling Montecatini to become part of the Italian national cartel, and a measure of control was set up for a time over the American independent sulfur movement in export trade by Sulfur Export Corp. finding a market for independent tonnage in its quota, in consideration of which American independents were to observe cartel prices. There were no such definite arrangements with American

independents after 1935, but the latter appear to have found it advantageous to fol-
low the price leadership of the cartel in sales they made in Europe.

When Rio Tinto Co. of Spain and Mason and Barry of Portugal, licensees
under Orkla patents, competed for the Spanish market and Rio Tinto threatened to
export to France, Sulfur Export Corp. exerted its influence to prevent such invasion of other markets.

In support of a cartel of French sulfur grinders who purchased their requirements from the cartel or from other sources that followed the cartel's price leadership, the main cartel participants undertook to bolster the French cartel's weak
control over the grinding industry by refusing to sell to particular French grinders
at destinations where new grinding capacity was being built.

The obvious international effects of activities of these types were to eliminate
competition in the foreign markets, enhance and maintain prices, retard the development of low-cost recovery of by-product sulfur through Norwegian processes,
accelerate the exhaustion of the world's richest natural sulfur, and compel foreign
market consumers to pay high prices in order to enhance and maintain the profits
of natural sulfur producers.

The domestic and export segments of the American sulfur industry are inseparable in interest. From an economic standpoint, the distribution and pricing activities of Sulfur Export Corp. have a natural relationship to the production,
distribution, and pricing activities of its individual producing members. Sulfur
Export Corp.'s international agreements have distinctly eliminated and restrained
competition abroad.

Its policy of tolerance toward the two smaller nonmember producers who
have been permitted, and upon occasion even assisted, to find foreign outlets far
tonnages, which additional tonnages, if sold in the domestic market, might have
had a depressing effect thereon, has tended to forestall and ameliorate competition
at home. The independents, while not always strictly observing Sulfur Export
Corp.'s prices abroad, or its members' domestic prices at home have nevertheless
followed quite closely the price and distribution leadership of their larger competitors in both markets because it was to their advantage to do so. To do otherwise
would not have been good business. The report of the Canadian Commissioner of
its Combines Act reports similar economic effects of the cartel agreements on sulfur distribution in Canada, one of the markets not included under the cartel agreements.

From 1926 through 1931, during the first three years of which there were
only two American producers it appears that the domestic price level yielded substantial profits. Price protection afforded the Italians maintained export prices sub-
stantially higher than domestic prices, and still further enhanced the profits of the
American producers. Quota restrictions under the cartel agreement, however,

restricted the quantity of sulfur that could be sold at the higher export prices. The fact that domestic prices were lower than export prices would seem to indicate that, especially prior to 1933, the two dominant companies, after selling abroad their permitted quantities under the cartel agreement at the higher cartel price, so priced their product at home as to retain the maximum share of the domestic market in competition with pyrites and by-product sulfur. Such pricing would be especially effective in retaining the tonnage consumed by the growing sulfuric acid industry.

The continued prevalence of this price difference in favor of sales for export during the late 1920's produced a situation in which it became profitable for American consumer-purchasers to buy in the United States and divert tonnage for export. Manufacturers of ground sulfur likewise found it advantageous to export to foreign markets where ground sulfur prices were based on cartel maintained crude sulfur prices. Since Sulfur Export Corp. was obligated under its cartel agreements to absorb such exports in its quota, a definite policy of direct selling only to domestic users, or to those intermediate handlers who would cooperate to preserve foreign cartel controls, was adopted both at home and abroad. A formal agreement between the two company members of Sulfur Export Corp., in 1929, to limit domestic sales to the needs of domestic users was abandoned when the Federal Trade Commission pointed out the objectionable effects of such an agreement on domestic trade. Restraints abroad necessitated restraints at home in order to determine not only who might engage in export trade in sulfur, but the conditions of trading.

The operations of all four producers constituting the American sulfur industry have been highly profitable, and the indications are that Sulfur Export Corp.'s foreign cartel agreements have added to the profitability of the United States industry even with respect to the non-member companies.

FTC – 1947, pages 82, 83

After formal hearings, the Commission issued recommendations for the readjustment of the business of Sulfur Export Corp. (Docket 202-6) on February 7, 1947, as follows:

1. That Sulfur Export Corp. refrain in the future from formulating, promoting or participating in any plan, program or agreement whereby either or any of the following described provisions, or provisions of similar purport or effect, are continued, entered into or effectuated, to wit:

(a), Provisions such as those in the agreement between Sulfur Export Corp. and the Ufficio per la Vendita Dello Zolfo Italiano, whereby said Sulfur Export Corporation bound itself to deduct from its tonnage quota of shipments of American sulfur for export, certain shipments of sulfur from the United States made by or through American producers not stockholders or members of said Corporation;

(b) Provisions such as those in the agreement between Sulfur Export Corporation and the Ufficio per la Vendita Dello Zolfo Italiano, whereby the latter was guaranteed the right to sell a specified minimum tonnage of sulfur in a certain designated period, on a priority basis over and above the tonnage of sulfur to be sold by Sulfur Export Corporation in said territory during the same period;

(c) Provisions such as those in the agreement between Sulfur Export Corp. and the Ufficio per la Vendita Dello Zolfo Italiano, requiring that shipments of manufactured sulfur from the United States made by or through American exporters be deducted from the tonnage quota of export shipments of crude sulfur made by Sulfur Export Corporation;

(d) Provisions such as those in the agreement between Sulfur Export Corp. and the Ufficio per la Vendita Dello Zolfo Italiano, requiring that the parties to such agreement were to maintain the status quo in the manufactured sulfur industry in the trade territories to which said agreement applied and to do nothing which would encourage any alteration in the competitive trade situation in said industry in said trade areas.

2. That Sulfur Export Corp. refrain in the future from formulating, promoting or participating in any plan, program or agreement such as that provided in the agreement with Orkla-Grube Aktlebolag, that Sulfur Export Corp. Shall acquire or control or participate in the acquisition or control of any share in patents or processes useful for or capable of being used in connection with the production of sulfur for commercial purposes, and that said Corporation in the future refrain from so obligating itself, financially or otherwise, in any such understanding or agreement.

3. That Sulfur Export Corp. in the future refrain from entering into any understanding or agreement with American producers of sulfur who are not regularly admitted and recognized members of said Corporation, whereby said producers or Sulexco agree not to sell sulfur in certain foreign markets, or to sell only at agreed or noncompetitive prices and terms, or to refrain from competing with each other in export trade in sulfur.

4. That Sulfur Export Corp. in the future cease and desist from selling, banding, marketing or disposing of sulfur for the account of or belonging to any American producer who is not a regularly admitted and recognized member of the Sulfur Export Corp.

5. That Sulfur Export Corp. in the future seasonably file with the Commission all information required by the Export Trade Act to be flied annually, and furnish all

information and documentary evidence requested or required by the Commission, pursuant to said Act, whether called for any report forms, by questionnaires or communications, by personal visitation or otherwise.

It is ordered by the Commission that Sulfur Export Corp. file with the Commission within 30 days hereof a report stating whether it has elected to comply with the above recommendations, and if so, the manner in which it has so complied. The sulfur association reported compliance and has continued to operate.

Source: **Annual Report of the Federal Trade Commission for the Fiscal Year Ended June 30, 1947**, p. 21 to 24

FTC – 1947, page 138

Sulfur Industry (F. T. C.) --In its report to Congress on *The Sulfur Industry and International Cartels* (6/16/47), the Commission stated that the operations of all four producers constituting the American sulfur industry generally have been highly profitable, and that the indications are that foreign cartel agreements entered Into by Sulfur Export Corp., an export association organized under the Webb-Pomerene Law, have added to the profitability of the U.S. industry. On 2/7/47, after hearings, the Commission recommended that Sulfur Export readjust its business to conform to law (see pp.21 and 82).

GLOBAL SULFUR PRODUCTION ('000 TONNES)

Table III.1. Sulfur production from 1900 to 1925

Country	1900	1905	1910	1915	1920	1925
U.S.	3	220	247	521	1,255	1,409
Chile	2	3	4	10	13	9
Italy	536	560	424	352	259	286
Spain	1	1	4	10	12	8
Greece	-	-	-	-	-	2
Japan	14	24	43	71	39	47
Global	555	808	721	963	1,579	1,732

Table III.2. Sulfur production from 1930 to 1950

Country	1930	1935	1940	1945	1950
U.S.	2,559	1,633	2,736	3,778	5,336
Canada	0	0	40	0	0
Mexico	-	3	-	7	11
Chile	18	20	32	29	22
Bolivia	-	4	4	1	4
Argentina	-	7	-	9	8
Peru	-	2	1	1	2
Columbia	-	-	-	-	1
Italy	364	307	325	74	210
France	-	-	-	3	6
Germany	11	34	-	-	-
Spain	12	31	4	5	7
Portugal	-	9	10	-	-
Greece	-	0	2	-	-
Norway	2	65	-	-	-
Sweden	-	10	-	-	14

Table III.2. Sulfur production from 1930 to 1950

Country	1930	1935	1940	1945	1950
Turkey	-	1	3	4	6
Israel	-	1	1	-	-
Japan	76	162	192	37	91
China	-	5	-	-	-
Taiwan	-	1	-	-	3
Global	3,000	2,700	3,500	4,300	5,700

Table III.3. Sulfur production from 1955 to 1975

Country	1955	1960	1965	1970	1975
U.S.	6,199	5,804	7,331	8,531	10,344
Canada	24	453	1,700	4,242	6,469
Mexico	506	1,326	1,560	1,359	2,130
Aruba	29	40	30	-	86
Bahamas	-	-	-	-	10
Trinidad	5	5	4	4	49
Cuba	-	-	-	-	8
Chile	56	31	35	107	21
Bolivia	4	1	9	16	22
Argentina	18	39	23	39	32
Brazil	-	-	5	9	19
Uruguay	-	-	-	-	2
Venezuela	-	-	-	-	81
Peru	-	-	-	-	16
Columbia	5	9	18	33	31
Ecuador	-	-	-	6	8
Italy	182	122	96	126	51
France	3	778	1,497	1,706	1,962
Germany	164	193	198	280	1,039
U.K.	46	62	48	36	60
Spain	7	40	43	5	3
Portugal	15	11	10	3	2
Austria	-	-		3	16
Switzerland	-	-	-	-	2

Table III.3. Sulfur production from 1955 to 1975

Country	1955	1960	1965	1970	1975
Netherlands	7	30	26	32	65
Belgium	-	-	3	10	193
Greece	4	-	-	-	3
Norway	99	71	0	0	5
Sweden	28	39	21	5	12
Finland	-	-	73	113	15
Denmark	-	-	-	-	8
Czechoslovakia	-	-	-	-	70
Poland	-	26	424	2,641	4,721
Romania	-	-	-	-	89
Hungary	-	-	4	3	9
Bulgaria	1	5	10	5	76
Yugoslavia	-	-	-	-	5
Russia	360	1,010	1,407	1,574	2,845
Turkey	11	17	4	26	79
Israel	-	-	-	8	10
Syria	-	-	-	-	1
Kuwait	-	-	-	47	54
Qatar	-	-	-	-	-
Abu Dhabi (UAE)	-	-	-	-	-
Bahrain	-	-	-	-	24
Saudi Arabia	-	-	-	5	3
Iran	18	20	20	407	487
Iraq	-	-	-	-	699
Egypt	4	6	4	1	10
Libya	-	-	-	-	20
Algeria	-	-	-	16	10
Zimbabwe	-	-	-	-	-
South Africa	-	-	7	16	25
Pakistan	-	-	-	-	13
India	-	-	-	-	6
Japan	200	251	246	337	775
Korea	-	-	-	-	10
China	-	250	246	246	167
Taiwan	5	7	6	10	7

Table III.3. Sulfur production from 1955 to 1975

Country	1955	1960	1965	1970	1975
Indonesia	-	-	4	1	4
Philippines	4	-	-	-	-
Singapore	-	-	-	1	6
Thailand	-	-	-	-	1
Australia	-	-	-	-	9
New Zealand	-	-	-	-	-
Global	8,120	10,400	15,286	22,162	33,484

Table III.4. Sulfur production from 1980 to 2000

Country	1980	1985	1990	1995	2000
U.S.	10,463	10,324	10,262	10,400	9,490
Canada	6,174	5,872	5,405	7,935	8,680
Mexico	2,102	2,020	2,120	882	851
Aruba	91	25	60	-	-
Bahamas	5	1	-	-	-
Trinidad	57	-	5	-	-
Cuba	8	5	7	-	-
Chile	88	79	29	-	-
Bolivia	11	3	2	-	-
Argentina	-	-	-	-	-
Brazil	131	59	99	-	-
Uruguay	8	2	2	-	-
Venezuela	85	88	125	180	450
Peru	20	68	66	-	-
Columbia	27	51	40	-	-
Ecuador	14	14	14	-	-
Italy	23	1	297	340	490
France	2,216	1,723	1,050	1,170	1,110
Germany	1,477	1,779	1,175	1,090	1,753
U.K.	84	80	135	-	-
Spain	15	9	151	77	116
Portugal	2	5	3	-	-
Austria	19	24	6	-	-

Table III.4. Sulfur production from 1980 to 2000

Country	1980	1985	1990	1995	2000
Switzerland	3	3	4	-	-
Netherlands	52	250	285	317	428
Belgium	270	260	310	347	410
Greece	4	135	140	-	-
Norway	6	10	15	-	-
Sweden	37	23	40	-	-
Finland	30	45	42	38	46
Denmark	8	7	12	-	-
Czechoslovakia	15	18	46	-	-
Poland	5,215	5,013	4,690	2,450	1,552
Romania	140	150	310	-	-
Hungary	9	9	9	-	-
Bulgaria	70	53	60	-	-
Yugoslavia	5	3	3	-	
Russia	5,240	5,084	5,925	3,390	5,550
Kazakhstan	see Russia	see Russia	see Russia	255	1,200
Uzbekistan	see Russia	see Russia	see Russia	320	-
Turkey	93	124	41	-	-
Israel	10	25	64	-	-
Syria	5	35	30	-	-
Kuwait	120	238	300	559	512
Qatar	-	37	52	-	-
Abu Dhabi (UAE)	-	105	90	257	1,120
Bahrain	33	36	5	-	-
Oman	461	1,100	30	-	-
Saudi Arabia	220	180	1,435	2,400	2,101
Iran	740	570	635	840	963
Iraq	3	7	1,180	475	-
Egypt	22	14	8	-	-
Libya	14	20	14	-	-
Algeria	5	5	20	-	-
Zimbabwe	25	100	5	-	-
South Africa	15	27	210	233	-
Pakistan	5	-	25	-	-
India	1,173	1,044	10	-	376

Table III.4. Sulfur production from 1980 to 2000

Country	1980	1985	1990	1995	2000
Japan	66	65	1,268	1,680	2,072
Korea	500	700	65	200	600
China	8	43	970	1,100	2,190
Taiwan	-	4	96	-	-
Indonesia	-	-	4	-	-
Singapore	12	50	65	-	-
Australia	13	12	70	35	30
New Zealand	-	1	3	-	-
Global	38,321	38,325	40,120	37,630	43,250

(*) volumes may contain by-product sulfuric acid, as some sources are unspecified

X

Xanthus 7

Z

Zeus 5
zinc smelter 114
zolfare 45